Spread and Growth Tectonics:
the Eocene Transition

Karl W. Luckert

a Triplehood Publication

distributed by Lightning Source

Second Printing

"Spread and Growth Tectonics: the Eocene Transition"
is the revised enlarged edition of *Planet Earth Expanding and the
Eocene Tectonic Event,* Portland. Copyright ® 1999.
Enlarged Edition, March 2016. Second Printing, May 2016.
Karl W. Luckert. All rights reserved

Library of Congress Cataloging-in-Publication Data

Luckert, Karl W., 1934—

geology, oceanography, tectonics.
Spread and Growth Tectonics: The Eocene Transition, includes
preface, epilogue, bibliographical references, and index;
pages: i-xiv, 1 - 158.

ISBN   978-0-9839072-6-8   Paperback

Plate Tectonics—Spread and Growth.
Expansion Tectonics.
Continents, Earthquakes, Mountain formation,
Oceans, Rifts, Trenches.
From the Jurassic through the Eocene

Contents

Illustrations

Fig. 1. United States Geological Survey—a World Map of Tectonic Plates.

Fig. 2. Expansion sequence of the oceans, from *Planet Earth Expanding and the Eocene Tectonic Event,* 1999; revised 2016.

Fig. 3. Drawing after Alfred Wegener, Die Entstehung der Kontinente und Ozeane, 1915/1929.

Fig. 4. NOAA topographic hemisphere, showing continental cohesion around the North Pole.

Fig. 5. NOAA, cylindrical-topographical world map.

Fig. 6. Isochrone globe, showing the southern hemisphere with the Antarctica Plate. On this globe, painted by the author, the Paleocene is indicated in a lighter tone than the Eocene.

Fig. 7. Earth Expansion in three stages. Page 68 of the author's essay in Theophrastus 1996.

Fig. 8. GSA Geologic Timescale Excerpt, from: http://geo-maps.wr.usgs.gov/parks/gtime/index.html.

Fig. 9. Experiment 1979-a. Putty slab on an expanding balloon, showing top view of Flanging, with "tensile folding" in the syncline. In 2002 it became obvious that the cracks and folds at the underside of such flanges were significant for future mountain building.

Fig. 10. Experiment 1979-b. Putty slab on an expanding balloon, held a little drier. It was loosened from the expanding balloon. The flange was left behind in the ocean as an island chain. Compare with East Asian island chains.

Fig. 11. Experiment 2002-a. Igneous mountain formation along the underside of a continental lithosphere.

Fig. 12 . Experiment 2002-b. Igneous dome formation along the underside of a less sticky lithosphere.

Fig. 13. Balloon skin cutouts similar to Africa and South Asia, resting on an expanding balloon; illustrating potential amounts of squeeze-materials accrued over time, to be squeezed sideways as the crust settles onto the flattening curvature of the sphere. The triangular flap, resembling India, keeps continental bulges inland.

Fig. 28: Approximate distances of movement during the Eocene Transition; on a present-day NOAA topo-map of the southern hemisphere. From its previous position (Figure 25), Australia was pulled away, for Antarctica to swing in. Then Australia with New Zealand, and Upper Cretaceous ocean floor, could drift east into vacated space. Thereafter Australia began to ricochet westward.

Fig. 29. NOAA Isochrone hemisphere, shows Upper Cretaceous floors from the South Pacific, at the back of Antarctica, subsequently burned through from a rifting heat below. Black lines were added by the author.

Fig. 30. Dora and Karl visiting the Great Bight.

Fig. 31. Rest-stop along the Ring of Fire.

Preface and Perspective

Rifts tore Continents—made room for Ocean Floor to SPREAD.
Continental shores GREW Ocean Floor to heal the Rift between them.
Ocean Floors will spread where Continents are touching, where Earth
encrusts her grainy Passion, Blood, and Glow.

Why not call it "Plate" Tectonics?

Many conflicts in science, culture and religion are consequences of misunderstandings, imprecise naming and ambiguously focused perspectives. From its inception in the 1950s and
1960s, the science of Plate Tectonics has been—as all human enter-
prises sooner or later are wont to be—haunted by ambiguities of
its own making. Human languages cannot communicate precision with words that minds, aspirations and ambitions, cannot
fully digest.

Here is what we have found, for a starter, summarized in
Encyclopedia Britannica: "The Earth's tectonics plates theory is deal-
ing with the dynamics of Earth's outer shell, the lithosphere... providing a uniform context for understanding mountain building
processes, volcanoes, and earthquakes, as well as understanding
the evolution of the Earth's surface and reconstructing its past continental and oceanic configurations.... According to the theory,
Earth has a rigid outer layer, known as the lithosphere, which
typically is about 100 km (60 miles) thick and overlies a plastic
layer called the asthenosphere. The lithosphere is broken up into
about a dozen large plates and several smaller."

We will underline the Alfred Wegener contribution of Continental Drift with a sentence from the *Merriam-Webster Dictionary:* "The lithosphere of the earth is divided into a small number of plates which float on and travel independently *[sic]* over the mantle and much of the earth's seismic activity occurs at the boundaries of these plates."

The designation of "plates" in Plate Tectonics was intended to combine Wegener's *"wandernde Kontinente"* with the discovery of an ocean-spreading process along mid-ocean rifts. This contrived synthesis immediately engendered a lopsided perception. The cartographers, who began depicting magnetic-reversal stripes as isochrones—similar to the method of tree-ring dating—gave us a nearly seamless, though sequentially "relative" ocean floor chronology. Scientists labored to define ocean floor plates by way of first establishing precise contours for the plate boundaries.

However, the insistence on precision that was essential for cartographers has shifted ocean floor explorations almost unnoticed toward the presence of "plates"—even where the plates of crust looked more like mud flows. Plates that had definable contours were in demand. They were sought and found. Henry W. Menard, the director of explorations in the Pacific area, had second thoughts. He suggested that mid-ocean rifts and ridges could be ephemeral features. By implication, this would render tectonic plates to be temporary phenomena as well. Menard's caution did not stop the cartographers. And this was a good thing for Earth science, where explorers subsequently hoped to rely on the maps they made. But nonetheless, Menard's concerns have not gotten answered.

The discovery of spreading rifts, and a renewed interest in Alfred Wegener's "wandering continents," excited Earth scientists. Had more time been spent on conceptualizing empirical "plate" data, we probably would have recognized more quickly

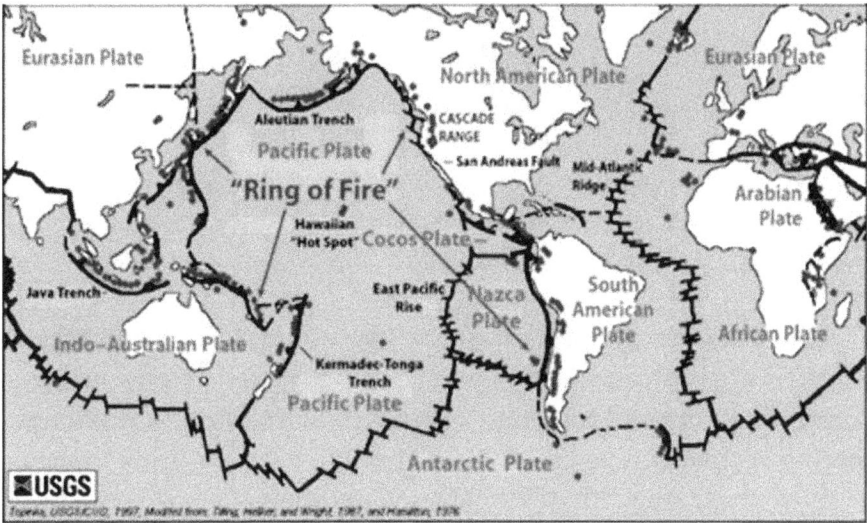

Fig. 1. United States Geological Survey—a World Map of Tectonic Plates

the general adhesion of ocean floors to continental boundaries. We would then not have strayed one-sidedly into asking only what spreading-rifts are doing. Upon discovering "plate formation" along mid-ocean rifts, the scientific mind—for the sake of control and understanding—immediately pondered the elimination of those plates. Military paradigms of subduction and destruction still hovered over those World War II ships, which scientists were given as tools to study ocean floors.

But ocean floors are not just simply "spreads," owned by oceans. How could ocean-water own the lithosphere that supports? The mafic floors could just as well be viewed as horizontal accretions along the shorelines of the older continents. The asthenosphere, which underlies all crusts, cannot be said to favor either land or ocean crust. The upper mantle supports and sustains both types of lithosphere.

With the kind of data available to him, Alfred Wegener could not yet have known that the continents he saw moving, were actually not independent wanderers. The paradigm of *"vagabund* continents" (this author's translation) which Wegener invoked, was basically wrong. On the other hand, had oceanographers in the 1960s taken more care to collapse Wegener's older theory to bare data, they could have extrapolated their own data better. Their new ocean floor data—the very ocean floor chronology that they constructed from isochrones—could have enabled a fresh and better beginning, without help from Alfred Wegener.

At the present moment in the evolution of Planet Earth, all continental segments of crust that belonged to the original Earth shell, are still touching at least one of their original neighboring segments. Therefore, there should be no enduring question of how these older than 200 Ma crustal pieces could have fitted on a planet poised for expansion. **Yes, the continents all have been rifting apart, and rifts have spread to become oceans.** Presentday mid-ocean rifts broke open first as narrow cracks in the lithosphere. They broke through the crusty shell and tore apart segments the size of the present continents.

Explorers, accustomed to negative thinking, who needed to focus on difficulties for submarine warfare along insufficiently known ocean floors, were under no academic pressure to look beyond the dips and ravines in the ocean crust, where submarines could hide. For bonus satisfaction, it was nevertheless interesting to discover how rifts were active and how ocean floors were getting spread.

But then, any scientific seafarer, who has the sensitivity to become enthralled by a Spreading Rift, just as easily could have become fascinated by coastlines and shores of continental crusts from which their ships sailed and to which those ships—though not necessarily the hearts of fervent explorers—were scheduled to return.

An ocean floor explorer could have noticed how continental felsic crust, resting on the same asthenosphere as did the mafic crust, might all along have been adding ocean floor by accretion, along its coast-lines. It would have been the same process. Why should all credit for ocean floor formation be ascribed to a hot rift or a gap? **The continents on Planet Earth, indeed, were torn apart while all along new lithosphere was being created.** But it was the asthenosphere, welling up, that furnished the lifeblood of creation. It was not the lithosphere that would do rifting by itself. The difference between continental and oceanic lithosphere ranges between felsic and mafic, all of which are likely being determined by thickness, depth, age, and the caresses that either water or the atmosphere could have added in the form of cooling.

Why not call it "Expansion" Tectonics?

In 1996, at Missouri State University, when in response to the worldwide proliferation of "Plate Tectonics" theory I built my first website, *kwluckert.com*, I named my perspective "Expansion Tectonics." A number of people adopted this label while, with some hesitation I began to neglect it. In the enterprise of university teaching, straight-forward antitheses are seldom appreciated for what they contribute. A straight antithesis to mainstream Steady-size Earth Tectonics, in those days, simply communicated opposition. In the refined climates of higher learning, single-step revisions seldom evoke anything better than irate attitudes. Nobody benefits from these.

We must seek a wide-angle perspective. Magnetic birthmarks of 200 Ma of spreading were found imprinted in the ocean floor crust, mirrored at right angles from the rift. A historical event, a consequence of the extensive reassignment of warships from World War II made this discovery possible. These vessels were placed in the service of oceanographic depth-sounding, scooping and drilling. As ocean plates continued to "spread" by hot lava, they can also be said to have "grown" from continents toward their far rifting edges—like fingernails grow to extend the reach of a hand. Continents reached and touched their parting neighbors with new ocean floor.

How about both—"Spread and Growth" Tectonics?

The word is out: Oceans are spreading along an only Mid-ocean Rift which splits and spreads all oceans. Continental shores are growing oceanic lithosphere along that same rift. What looks like "spreading" along oceanic rifts appears to be "growing" onto continental shores—onto shores which, over the past 200 Ma were lowered below sea-level to grow mafic crust.

Most ocean rifts appear to have moved sideways very little. On account of prior cohesion and tension among continents, of different speeds by which the flanks of rifts could widen, oceans, especially the Pacific, grew unevenly. A few continents "rifted" farther away from others, and some partial severances of lithosphere from upper asthenosphere occurred. But no continent detached completely from all its neighbors to go "wandering."

Five dimensions of thought are required, nowadays, to harmonize Plate Tectonics data with other systems of explanation. First there are the Three Dimensions of Space—breadth, length, and height. The fourth dimension represents changes that happen in the Flow of Time. For ordinary tasks and scientific orientation these four dimensions suffice. But as pertains to Plate Tectonics, where worldviews from American-English empiricism and German poetic naturalism had been converging a century earlier, **the Fifth Dimension which pertains to the dialectics of worldviews, history and language, has gotten engaged as well.**

*   *   *

During the 1950s, when this German-born wide-eyed student set out to find his trail into American academic empiricism—a mental attitude at the time still unfamiliar to him—some of America's stalwart Earth scientists, instigators of the Plate Tectonics Revolution, were seeking the grail of scientific Earth-theory in Germany. Thirty years after Alfred Wegener went missing in Greenland, his theory of Continental Drift—of *wandernde Kontinente*—suddenly was given commodity status. The new costumers were highly successful American explorers of ocean floor. Prior to their pilgrimage to Germany they were able to enlarge the data-base concerning ocean floors beyond anyone's expectations, including their own. They have discovered mid-ocean ridges and the implied continuous Rift that was enlarging all the Planet's ocean floors. They discovered ocean-spreading and de-coded magnetic anomalies which the process of spreading has imprinted into the seafloor. They enabled themselves to determine the relative ages of stripes and patches of ocean floor. In addition they succeeded with improving their topo-graphical maps. By now some great refinements of satellite altimetry have been added as well.

Accolades given by representatives of Plate Tectonics science—to Alfred Wegener as their first founder—are a nice gesture toward German-born scientists in the English-speak-ing world. They are, however, a gesture that has bequeathed ambivalent consequences. **"Continental Drift" and *"wandernde Kontinente"* do not complement the worldwide Mid-ocean Spreading Rift which Plate Tectonics explorers have discovered during the 1960s.** While half a century earlier Wegener has argued for Continental Drift, he may indeed have alluded to the Atlantic separation as "rifting." Nevertheless, having been granted first founder status on the basis of "Continental Drift" seems to have impeded progress in Plate Tectonics reasoning.

The landmasses which have gotten torn, by rifts that have widened and spread oceans, have nothing in common with "wandering" or "drifting" continents. All the seven continents on Planet Earth are still today touching at least one of their original neighbors. Therefore, tracing continents tectonically to their original positions is not all that difficult. However, what is difficult and even impossible to accom-modate to Earth science are imaginary "drifters."

Continental Drift theory has misled a majority of explorers into postulating "ocean floor subduction" and "convection currents in the mantle"—essentially to the end, that the continents, which an acknowledged founder has defined, would be enabled to continue wandering in his honor. Such honorary hypotheses do not help explain the known data. They only assist in the theoretical disposal of imagined surplus ocean floor; they clear the road for imaginary *Vagabund* continents. They enable scientists to think of continents as having wandered where none of them has ever gotten to.

Of course, Alfred Wegener wrote at a time before anyone knew about the Mid-ocean Spreading Rift or about its habit of leaving magnetic footprints in ocean floor lava. Greater help than Wegener's hypothesis to Plate Tectonic science, therefore, is the new ocean floor chronology which the oceanographers themselves have created. The new Isochrone maps provide the best information about where on Earth, in the course of 200 million years, our seven continents may have been located in relation to one another.

To mention a little more about the aforementioned "Fifth Dimension" of research, pertaining to cultural relativity and context, I have in this Second Printing enlarged the Epilogue (pp. 147-150) with links to interdisciplinary causalities. Most Earth scientists probably would prefer not to pay attention to the inescapable "drift" that happens to important words in all human languages. But for the benefit of future generations of scientists it should be said, that all accomplishments of honing brittle scientific jargon, will require continuous maintainance and repair—very soon after stabilization has been achieved, or not less than once in every generation. The verbs "to wander," "to drift," to "rift," "to spread" and "to grow" have presently ended up on our workbench.

Part One

Spread and Growth

Some highlights in this book are printed in Berlin Sans FB typeface. Passages so emphasized are intended as a tool for quick review. The Subject Index, at the end of the volume, may open a next larger context.

1

Introduction to *Pangaea* the All-land

The Pangaea of Alfred Wegener

Once upon a time, approximately 300 Mya, a super-continent assembled on Planet Earth. Scholars who knew Greek named it "Pangaea." In English this name means "All-Land." The name seems to fit, because surrounding the All-Land was the All-Sea, Panthalassa, which could conveniently be distinguished from the All-Land. Approximately 280 Mya gaps appeared among the Planet's continental crusts. By 200 Mya, the cracks had become wide gaps; and after another 100 Ma, huge continental segments began to break away from the total All-Land. These segments drifted. They began to wander and to arrange themselves on the surface of Earth where they can be seen and explored today.

This is a summary of the story, as it is told around the world today — from elementary schools, up to the high plains of scientific institutes and universities. Recitations of this story are being sponsored by several of the Earth sciences, as well as by paleontology. Thanks to these sciences, **with help from the newest ocean floor topo- and isochrone-maps, I will attempt here to re-examine the surface areas of Pangaea as well as Panthalassa.** Such a review, today, is no longer an impossible undertaking, because, in our days, maps of lands and oceans can be found on the Internet. Everyone can continue at the point where I am leaving off.

3

All one needs to do is to enter the acronym NOAA (National Oceanic and Atmospheric Administration) into a search engine of choice.

It is not my intention to continue building on established scientific hypotheses, even though far and wide some of these have been promoted to the status of theory. For the time being I am concerned about understanding some actual conditions in our planet's crust, how this crust has become observable under water almost as easily as under air—which means, as presented superficially on maps. Whether at any time the story about Pangaea has represented truth, or merely true fantasy, does also not interest me at this moment. Instead, I wish to let the new land- and ocean maps provide fresh insights regarding magnetic traces and geological circumstances in Pangaea and Panthalassa .

"Pangaea" Review for Hamburg—2015

On April 26, 2015, I received an invitation to come to Hamburg to present there, on June 12th, two lectures of unspecified lengths, at the *Studt/Gärtner wissenschaftlich-politischen Salon*. My presentations were to be delivered to a select audience and were to cover (1) the history of religions and (2) my theory on the expansion of Planet Earth. I owe the opportunity and the invitation to Professor Karl-Heinz Jacob (Berlin), who has recommended me to friends and colleagues. I accepted the invitation to visit Hamburg and, thereby obligated myself to begin revising some of my earlier publications on Earth Expansion.

The booklet, *Planet Earth Expanding and the Eocene Tectonic Event* (Lufa Studio/Triplehood, 1999), was at the time sixteen years old. Not even with my best intentions would I have been able to promulgate my thoughts regarding the process of Earth expansion in that outdated form. Revisions needed to be made.

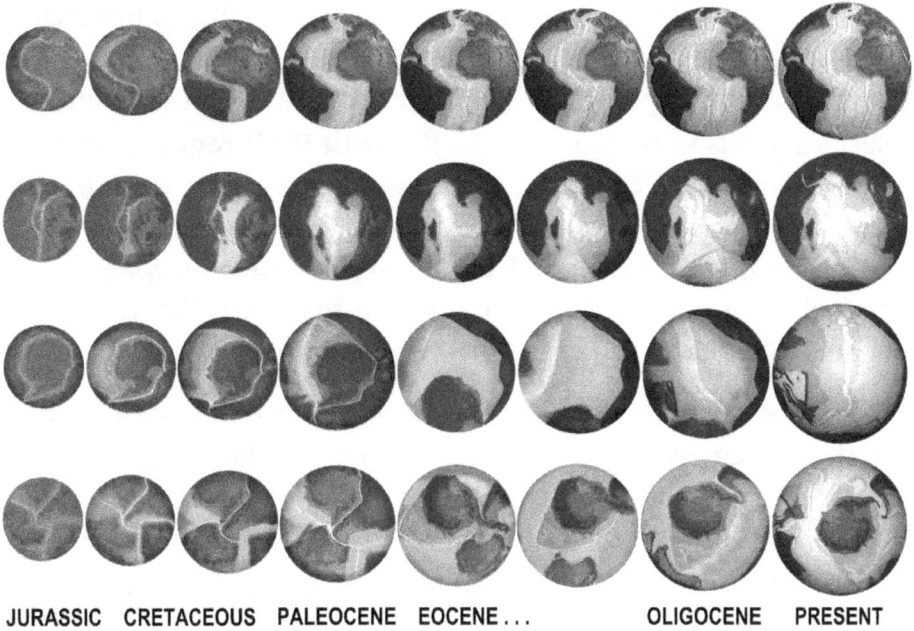

JURASSIC   CRETACEOUS   PALEOCENE   EOCENE...   OLIGOCENE   PRESENT

Fig. 2. Expansion sequence of the oceans, from *Planet Earth Expanding and the Eocene Tectonic Event,* 1999; slightly revised 2015.

The 1999 booklet was published as a limited edition. It was distributed at geological conventions and placed on my website www.triplehood.com. The extent of its reception is not well known. The subject matter of Earth Expansion surely has not yet become the popular topic that it will be a century from now. Whether the content of this booklet ever was fully understood by anyone, this also remains an open question. It is unknown whether any reader has taken the necessary time to think intensely about the recently discovered magnetic polarity reversals along sea floors, as illustrated on isochrone maps and published to reveal relative ages of the ocean floors. These maps were distributed worldwide by 1988, in the *UNESCO Geological World Atlas.* Newer and better maps of the ocean floors have since provided greater topographical detail. For this edition, some spheres in Figure 2, showing the movement of Antarctica during the Eocene, can now be focused a little better.

With the present oversupply of printed and digital materials, one sometimes wonders whether anybody will ever get as far as to examine hypothetical sequences of continental separations, seriously based on isochrones. **The task itself requires intense five-dimensional thinking: to engage oneself with data in the three dimensions of space, in the flow-dimension of time, as well as in the fifth dimension of the history of world perspectives**— namely, considering how hitherto the first four dimensions of space and time have been dealt with by human minds.[1]

These circumstances leave my 1999 booklet, in English, stranded as a neglected orphan and they present the challenge, that the 1999 English version, *Planet Earth Expanding and the Eocene Tectonic Event,* be revised in light of ideas presented in the newer 2015 *"Pangäa und Erdausdehnung, Review für Hamburg"* lecture.

The Continents according to Alfred Wegener

In 1915 Alfred Wegener published *Die Entstehung der Kontinente und Ozeane,* in which he showed how the continents had been separating from an assembly of lands that he named Pangaea (All-Land). There was general opposition to his ideas. But recent international progress in oceanography, with the discovery of the world-encompassing mid-ocean spreading rift and the feasibility of an ocean floor chronology, has generated fresh interest in the publications of Wegener.

Younger researchers in Plate Tectonics have combined Wegener's notion of "wandering continents" with the discovery of the global mid-ocean rift, of tectonic plates, of symmetric magnetic striping and the feasibility of an implied chronology. They synthesized this set of ideas by adding supportive hypotheses about "convection currents in the mantle" and the "subduction of ocean floors."

---

[1] See "Epilogue—the Fifth Dimension;" Tectonics of a Hiking Song, pp. 147ff

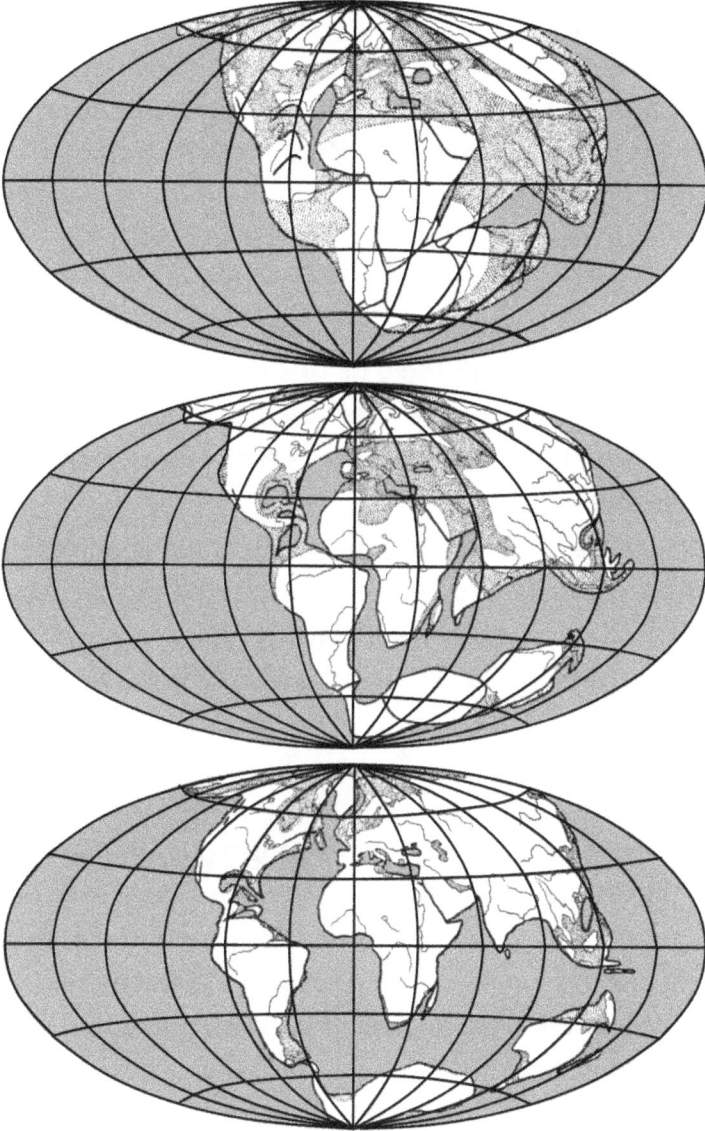

Fig. 3. Re-drawn after Alfred Wegener,
Die Entstehung der Kontinente und Ozeane, 1915/1929.

Caught up in euphoria about the new oceanic discoveries, under the aegis of natural sciences, it was generally not noticed that Wegener, as well as his latter-day followers, had all along been laboring under an ancient cloud of mythology. Of course, this fact alone does not prove that they were wrong. Surely, ancient mythology had some things right. But this circumstance should serve as an incentive for careful historical introspection, for the examination of contexts, and for common sense in the context of fashionable worldviews.

There are two mythical notions that haunt those who study tectonic plates in the shadow of Alfred Wegener. First there is the "central land," surrounded by the "world ocean," and second, there is the "middle sea"—das Mittelmeer. Wegener was under the spell of both.

Ancient Mesopotamian land dwellers envisioned their All-Land to be centrally situated. It was watered by rivers and surrounded by the boundless and mostly unknown world ocean. For merchants who crisscrossed the Mediterranean Sea and who hugged its shores, their homelands lay nicely arranged around this watery center. Their "middle sea" enlarged for them the former functional centrality of the Euphrates, Tigris, and Nile rivers. Farther out in this geography, the known lands were believed to be surrounded, still, by the world ocean of Near Eastern antiquity. **The complete geographical model, of the surrounding world ocean and of a sea in the middle, has gotten established in the psyche of Western peoples as an archetype.**

When Alfred Wegener contemplated the matching shorelines of the Atlantic Ocean, he brought together into his modern world model all the lands that seemed to have strayed from their original unity, the Pangaea. Wegener's idea of an All-Lands „Pangaea" was not necessarily a bad idea. It could have been nicely adjusted over time, had it not been skewed from the start by his other quasi-mythical concept of the remainder, the world ocean

Panthalassa, which he posited comfortably as outer wrapping around Pangaea. **For the Western mind of Wegener this meant that continental shores needed to match only along the "middle sea," which for Pangaea now happened to be the Atlantic. At the back of the globe the shore-lines of continents were permitted to slacken out loosely into the vast, unknown Panthalassa.** The happenstance, that shorelines along the Atlantic can be matched more easily than those of other oceans, seemed to endorse Wegener's mythic predilections.

Today, among latter-day Plate Tectonics scientists, the notion of a Pangaea-defined ocean-spreading process is being espoused mostly on the basis of the easily matched shorelines of the Atlantic. Even if only for the ease of fitting together these coastlines, the version of Wegener's Pangaea appears destined to abide with us for a while. In addition, some subduction-oriented theorists still continue to be haunted by the "middle sea" phantom. Among the oceans which now fill spaces among the scattered lands of Pangaea, some scientists still grant special status to the Atlantic—namely, with an exemption from the natural law which, supposedly, governs the subduction of seafloors in other oceans. Hardly any deep trenches can be found in the Atlantic. This ocean is therefore, generally and for the time being, given the freedom to spread without having to subduct its "surplus" floors.

Of course, from the perspective of a supposed natural process, this exemption does not withstand a first glance of logical reasoning. According to the new magnetism-based ocean floor chronology, the Atlantic, Indic, and Pacific feature similar quantities of epochal isochrone stripes or patches along their floors, which have all accumulated since Jurassic times. In order to avoid a fairer comparison, the Indic and Pacific have on that account gotten confounded with the somewhat still mysterious acclamation of "Panthalassa," with all its fog and antiquity.

An all-surrounding world ocean, kept close to the Wegener tradition, and endorsed by the central presence of an original Pangaea, surely can devour any amount of excess ocean floor. Why should an all-surrounding World Ocean not be able to do this?

The idea of a Pangaea and of a surrounding world ocean was long ago conceived with a Flat Earth worldview. Flatness also enables modern Pangaea dwellers to visualize and to arrange their geography quite easily on flat world maps. This expediency enables land dwellers to simplify the subject matter to fit—to neglect evidence from the flooded backside of their sphere. In spite of protestations to the contrary, the convenience of a two-dimensional geography is still preferred by many modern students of geology. Fully developed three-dimensional thinking, which in addition should be visualized in the flow-awareness of the dimension of time, cannot arbitrarily select just any one ocean to be either central or peripheral. One cannot just banish all the difficult "land questions" out into an ancient and all-surrounding Panthalassa. While Wegener's theory can be animated within a two-dimensional view, on a flat world map, animation of our Spread-and-Growth Tectonics requires several perspectives on hemispheres.

Meanwhile, even people who espouse an Earth Expansion hypothesis have been susceptible to ancient mythology. For instance, the goddess Tethys, anciently revered along the shores of Greece, has proven herself to be a very persistent divine lady among avant-garde Earth expansionists. After the famous spokesman for Earth Expansion, Warren S. Carey, reduced all oceans to zero on his pre-expansion Earth model, Tethys continued to haunt him under the guise of "Tethyan Shear," a process. It is a concept which I consider to be as unnecessary as was Alfred Wegener's "Panthalassa" earlier. Tethyan Shear phenomena can more easily be explained, as equatorial bulge-stretch-and-tear occurrences on a rotating and expanding globe.

Fig. 4. NOAA Topographic-hemisphere, showing continental cohesion around the North Pole.

All Pangaea Continents still hold together

The continental fragments of Pangaea, which Alfred Wegener has declared to be free wanderers, have not wandered away. They adhere together still today. All are attached at least at some point, still, to at least one of their original neighboring slabs of crust.

While South America and Australia appear completely severed from each other, they nevertheless have remained attached to their neighbors on opposite sides. Even Antarctica still hangs on, barely, by a thin stretch of messy shelf crust at the tip of South America. **The complete Pangaea curtain of continental shreds still hangs draped over the globe in three torn strips.** On a NOAA topographical hemisphere, this curtain can be seen fitted nicely over the North Pole (Figure 4). On a cylindrical world map (Figure 5) we notice how, in the course of 200 Ma, the continents around the globe were pulled and torn into three southward draping shreds.

Fig. 5. NOAA, cylindrical-topographical world map.

These strips of land appear even more contiguous if continental shelves and shallows, below sea level, are viewed together with the continental crust. For the purpose of rendering spreading and rifting processes more apparent, **we shall think of the three shredded continental strips as a "harness," as representing three half-suspender-straps. Later we discuss two of the half-straps as they continued to function as a "belt."**

A rotating planet tends to flatten at its poles. Centrifugal force along the equatorial circumference tends to bulge the crust and thereby creates tensions which cause cracks and crevices. Earlier geologists have referred to this weakened equatorial belt as "Tethyan Shear," or have even thought of it as the remains of an earlier Tethyan ocean. Stretched continental remnants and bent ridges are readily apparent. However, the process of circum-global shearing, as such, remains an instance of scientific or artistic speculation—probably a little of both.

As far as one can tell from the updated ocean floor chronology, the first deep ocean fissure broke through the crust already during the Triassic epoch, at a place that became the eastern Mediterranean Sea. An additional small rift opened up later in the Gulf of Mexico, during the Jurassic, and another small rift in the Arctic Ocean.

In the Jurassic period three major cracks, caused by general Earth expansion and aggravated by the Planet's centrifugal force, opened larger oceans which, eventually, severed three strips of continental lands from each other. Toward the south the three oceans were torn open.

Creeping slates and squeezed magmas, from below, could not fill all the widening cracks in the crust speedily, nor could they cool, harden and heal from above in short enough time. But thanks to the space that was being generated by the planet's sustained expansion, breaking forth through the upper mantle, these larger Jurassic cracks were able to break through, all the way to the surface of the crust. The crevices widened and their spaces proceeded to spread three huge oceans overhead.

Between these three oceans remained three tattered longitudinal strips of continental crust—three somewhat resilient straps at their undersides mysteriously hot-glued to the asthenosphere. We may think of them together as three elastic suspenders which constituted a "harness." Two of these straps we continue to refer to later, jointly, as comprising a single "belt." The three longitudinal straps of the continental "harness" were remnants from the Planet's former crust. They consisted of 1) Australia and Asia, 2) South America and North America and 3) Africa and Europe. **During the Jurassic period all three straps held together, still, at both of the planet's poles. Between the tattered stripes of continents spread the three large oceans.**

Over time, two great breaks happened at the straps of this "harness," very near each other in the original crust, at their southern juncture. The first break occurred early in the Lower Cretaceous and the second broke a hundred million years later, during the Eocene.

The Europe-Africa strap of the harness was broken off when southern Africa separated from South America, deep through the crust and tearing into the upper mantle. The Lower Cretaceous rift, which began to open the South Atlantic, tore an arc around the Cape of Africa and onward into the rift of the Indic Ocean. The first rift of the Indic had been spreading already since Jurassic times.

Eocene Events and Transition

After the breakaway of southern Africa from the toe of South America, global expansion continued to generate circumferential tension in the crust. But after Africa's breakaway, overall global expansion-tension was reduced from the three-strap harness to the two remaining straps that continued to hold together as a longitudinal "belt." **The harness of three straps was reduced to two straps which continued to hold tight from Austral-Asia along Asia, the North Pole and the Americas. The two straps continued to function as a single belt that longitudinally enwrapped and constricted the expanding planet.**

Both strap segments of this belt were severely strained along their equatorial mid-regions as well as in the Arctic. At these places they were elongated and nearly pulled apart. The Sunda Shelf, that is, Southeast Asia, was held below sea level over wide areas. At the other side of the globe, along Middle America, the other segment of the belt endured a similar fate of stretching, especially in the Panama region.

For the duration of nearly 100 Ma, prior to the continental belt's rupture during the Eocene, continental flanges were subjected to extreme tensile folding. This tensile folding has been partly responsible for the agitation and later rise of the Rocky

Mountains in North America and of the Andes in South America. It has lowered land levels not only along the equatorial stretch zones, but also has submerged other continental land to the level of sea shelves. Much of the Arctic area has become a shallow Sea. The Western Interior Seaway consisted of shallow waters that, longitudinally, flooded much of North America and lasted from the Jurassic well past the Cretaceous. It was uplifted and drained when, approximately during the Paleocene and Eocene (dates are not adequately established), the constrictive belt finally loosened and the craton of North America could rise and find its new natural levels.

But this last continental belt broke neither at the thinly stretched equatorial Southeast-Asia, nor in the Arctic or in Middle America. Antarctica, during the Triassic, was still part of the planetary land-crust and was still wrapped into a curved embrace by the Americas. But during another 100 Ma after Africa's breakaway, the global belt has gotten overstretched. The global belt was straightened by tension, and therewith the two Americas have expelled the round continent slowly from their horizontal embrace. They nudged it sideways, especially where the rounding Antarctic coastal rift (a continuation of the enduring Jurassic rift in the Northwest Pacific) had since the Lower Cretaceous been curving in, and cutting in from the north between Alaska and the Antarctic plate. The Americas, stretched as a portion of the global belt, under the general tension of Earth expansion, sidled the round continent westward. Antarctica was thereby sent on its way to turn southward, in a counter-clockwise motion (see Figures 26-28).

In the course of the Eocene Transition, the Bight of Australia and the Cape of South America were sliced apart in a curve as well. Antarctica accomplished this task by wedging its tail—that is, by using its tail as a "heel" or a "wedge." The Bight of

Australia and the variously bent tail of Antarctica, were both party to the impact scar that can be seen along the cape of South America. The place was scooped and torn into. With the curiously ripped landscapes, all three continents survived to tell their story of geological mangling, of long ago during the Eocene. The violence begun already during the Lower Cretaceous.

Antarctica, having begun to turn counter-clockwise in the north, slowly slid southward and pushed its curved tail, like a bent chisel, deeper along the Bight of Australia. The round continent finished carving the Great Bight of Australia by cutting the continent loose. After severing Australia from the Cape of South America, Antarctica continued its counter-clockwise swing and pushed its heel eastward to scoop the South Sandwich Islands into being. And then, while retreating west, it left the ridge it had scooped, and left a trench of tensile withdrawal lay in its wake. Antarctica pulled its "wedge" back west and continued its southward swing by severely bending that wedge in the process, like a "tail." The tail was bent sharply southward, so that the continent could swerve into its present position, to occupy the South Pole.

The Eocene Event has radically altered the Planet's geography. As if in retaliation, for having been ejected from the embrace of the Americas, Antarctica cut through that self-same asthenospheric longitudinal belt that itself has nudged the round continent westward in the first place (Figure 26). It severed the Cape of South America from the Great Bight of Australia. And all this probably happened between 43 and 42 Mya.

2

Spread and Rift Chronology

During one of his humble semi-public moments the great Sir Isaac Newton (1642-1727) opined:

> I do not know what I may appear to the world; but to myself I seem to have been only a boy playing on the seashore, and diverting myself in now and then finding a smoother pebble or a prettier shell than ordinary whilst the great ocean of truth lay all undiscovered before me....

Henry William Menard has quoted this passage at the opening to his fourth chapter, in his book, *The Ocean of Truth, a Personal History of Global Tectonics.*[2] He thereby has, in effect, acknowledged the origin of the title of his book. Menard's work was published by Princeton University Press, in 1986, the year of his death. Throughout he has made an effort at cultivating his image as that of a wise historian and impartial patriot of the scientific Plate Tectonics Revolution. At that, the choice of his book title reveals some of the bravado that inspired him and his compatriots. The discoveries that this scientific revolution claimed were not just a few smoother pebbles or prettier shells, but the whole great Ocean of Truth.

Of course, Menard merely made use of another man's metaphor. When Isaac Newton referred to a great ocean of truth, he surely had the universe in mind. Menard imploded Newton's metaphor and applied it to the oceans of the world which, in the

---

[2] Menard. H. W. *The Ocean of Truth: a Personal History of Global Tectonics.* Princeton: Princeton University Press, 1986.

course of the Plate Tectonics Revolution, and with the discovery of a continuous Earth-encircling spreading rift and ridges, had become a single world ocean. The specific portion of this world ocean, where Menard himself has searched and found most of his pebbles of truth, was the Pacific. William Wertenbaker, the biographer of William Maurice Ewing at the Lamont Geological Observatory, referred to Henry W. Menard, Ewing's counterpart at the Scripps Institution of Oceanography, as the one who almost single-handedly was discovering and explaining the seafloor geography of the Pacific.[3]

Since January 1997, when I began exhibiting my "Expansion Tectonics" hypothesis on my Missouri website <www.kwluckert.com> (now www.triplehood.com), I received occasional inquires from students who were assigned to write research papers on Plate Tectonics. Yes, occasionally I was tempted to guide them straightway to the Earth Expansion literature. But I did not do so. These students had to earn their grades in established academic environments. Therefore I recommended that they begin with Henry W. Menard, *The Ocean of Truth*, and work from that body of information outward. At least one graduate student wrote back, thanking me for the good advice. Her paper had turned out well.

Now that I am about to write my own essay, I feel obligated to heed my own advice. Start with Menard! A better insider history of the Plate Tectonics Revolution will probably never be written. Typical polemics against the present Plate Tectonics stampede, against the mindset which in some ways has closed itself off to any older or newer considerations—of which I have written occasional pages and usually torn up—definitely do seem out of bounds after meeting as fair a man as Henry William Menard.

---

[3] William Wertenbaker. *The Floor of the Sea, Maurice Ewing and the Search to Understand the Earth*. Little, Brown and Co., Boston, 1974, p. 176.

But no! Neither he nor any of the other heroes of the Plate Tectonics Revolution—the Ewing brothers, Heezen, Dietz, Revelle, Bullard, Raitt, Wilson, Fisher, Worzel, Hess, Vine, Matthews, Meinesz, Heirtzler, Pitman, Sykes, McKenzie, Morgan, and more—have touched bottom in Newton's "great ocean of truth." They explored only one type of ocean, the kind that holds saltwater. And certainly, they came to understand more of it than anyone had before they started. They found many hitherto unnoticed "smoother pebbles" and "prettier shells."

According to Menard, the problem of Continental Drift, as Alfred Wegener had defined it, has been solved by the scientists that gave us Plate Tectonics. To this author it looks as though Plate Tectonics has not solved, but actually modified Wegener's puzzle. They have proven the fact, indeed, that oceans do spread along a single continuous mid-ocean rift. They have not proven Wegener's supposition that continents do drift or wander. Their evidence implies that, while the present set of oceans has indeed been spreading, continents must have been distancing themselves from each other. If anyone asked specifics regarding the ocean spreading, the easy Atlantic was usually offered as an illustration.

The continents in Wegener's Pangaea (All-Lands), that touched the Atlantic, only appeared to have "drifted apart" as a result of obvious rifting and ocean floor spreading.

Drift is something that happens horizontally, at random, over a large area. On that account the idea has often been coddled within the limited context of two-dimensional space. A scientific determination of "drift," it seems, would as well depend on one's orientation in the third dimension. For example, tree branches that are being swayed horizontally, by the wind, are not considered to be drifting relative to the third dimension of height. In similar fashion, the continental crusts that became separated as a result of ocean floor spreading may not have been drifting all that much if the sphere itself was expanding and if continents remained connected similar to the mode in which branches are held together by the stem of a tree. Relative to our planet's core, the continents may simply have been rising. **By way of concentric terrella models Klaus Vogel has illustrated how three-dimensionally, and without exception, all continents have gotten separated by rising**.[4] In the case of most of our continents this appears approximately true.

Since 1979 I have, on my own terrella models, arranged the Jurassic surface of the planet somewhat differently, I personally do make a few exceptions in favor of differential horizontal movement.[5] **Uneven mantle expansion and unequal horizontal separations have resulted in the greater-than-average expansion of the Pacific and of the Southern Ocean.**

However, in the case of at least four continents there have been some extra horizontal movements, some kind of leaning or sliding away from their prior positions relative to the Planet's core. Australia has pulled away from South America, north and eastward, at the opposing side of the globe. Subsequently it has proceeded to adjust itself again westward. (All references to directions herein pertain to alignments relative to the present globe). From near the present South Pole, South America has pulled away from Australia possibly over a somewhat shorter distance than Australia appears to have moved.

At its southern portion, North America was twisted a little ways westward, by South America's impact against the Florida area crust. Then, the Antarctic Plate has managed to twist out of the eastern Pacific cavity, away from the larger ocean cavity where it was given shape. While these happenstances imply some

---

[4]Klaus Vogel, "The Expansion of the Earth, an Alternative Model to the Plate Tectonics Theory." *In Critical Aspects of the Plate Tectonics Theory*, II. 19-34. Athens, Greece: Theophrastus Publications, S.A., 1990.II.

[5] Karl W. Luckert, "A Unified Theory of Earth Expansion, Pacific Evacuation and Orogenesis," in *Theophrastus' Contributions to Advanced Studies in Geology*, 61-73. Athens, Greece: Theophrastus Publications, S.A., 1996.

amounts of horizontal "drift," they do not come anywhere close to supporting Wegener's *vagabund* paradigm of his "*wandernde Kontinente*" which, supposedly, always have been adrift in Panthalassa.

From the newer perspective, as it was demonstrated by the Plate Tectonics Revolution, ocean floor spreading is no more and no less than what its name implies — the spreading or widening of the ocean floors. This "ocean floor spreading" does not necessitate the "wandering" or "distance drifting" of continents, nor does it really consign or subduct patches of ocean floor to a fiery abyss in the upper mantle. The presence of Benioff earthquake zones and volcanoes along the famous Ring of Fire does not provide meaningful evidence in favor of convection currents in the mantle or in favor of ocean floors being subducted down into the mantle.[6]

Ocean floor spreading happens in rifts, along ridges in the Atlantic, Pacific, Arctic, Antarctic (Southern) as well as in the Indic oceans. Inasmuch as no one so far has shown that continental crusts are shrinking, the oceans cannot altogether be expanding at the expense of the continents. And sediments at the floors of the deep marginal trenches, where subduction of ocean floor crusts was supposed to happen, appeared peaceably undisturbed from the outset of the explorations. In addition, in spite of claims to the contrary, global seismic tomography has failed to show credible subducted slabs of ocean crust going down into the mantle.

---

[6] Volcanoes served as a model for Dante's notion of purgatory — a place for reforming and remodeling human souls. In Melanesia some of the volcanoes are proof of the campfires of deceased tribesmen, who in a different state of being dwell beneath. The subduction and recycling of ocean floor crust seems to belong to a similar category of mythology. Someone apparently has begun the habit of hiding intangible objects of faith (old ocean floors) in the magma fires of hell.

Judged solely on the basis of empirical evidence, it does not appear as though to this day ocean floor subduction has ever happened outside of a human mind. When it comes to the disposal of fictitious old ocean floors we should hold ourselves to the same standards of honesty which our scientists were able to maintain when they first learned about the creation of new ocean floor along mid-ocean ridges, while concurrently learning something new about magnetic reversals during lava-accretion.[7]

"Smoother pebbles" and "prettier shells" have indeed been found. They were polished as jewels and were inlaid to adorn the crown of Earth Science. **The crowning achievement of the Plate Tectonics Revolution was the discovery of ocean floor spreading and the new art of sketching ocean floor chronology, based on magnetic, paleontological and radiometric profiling.** The completed crown of the Plate Tectonics Revolution lies now displayed and published, as isochrones on ocean floor maps, in the form of a nearly seamless chronology. For the moment I am satisfied with summarizing how Menard himself has explained his revolution. It is always wise to let an insider have the first word:

After a decade, by 1960, the ad hoc explanations for individual groups of observations were no longer very satisfying. It was time to integrate all the new data with the old and to generate testable syntheses.... I shall group them in three classes: (1) "sequential" hypotheses in which ridges are of different ages and convection acts at different times; (2) "expansion" hypotheses in which crust is created but not destroyed by expansion

---

[7] This is not to say that proponents of Earth Expansion have, as a rule, faced new data more objectively. Some in our midst have gotten frustrated by the ocean floor chronology and have disallowed the magnetic record outright. The new ocean floor chronology has suited neither the preconceived notions of Plate Tectonics revolutionaries nor those of the Earth expansionists. Nevertheless, after solving the puzzle of the Eocene Tectonic Transition, objections launched from either side should appear less obligatory.

of the earth; and (3) "sea-floor spreading" hypotheses in which crust is created and destroyed by mantle convection. All the phenomena can act at once. The earth can expand while convection subducts some crust into the mantle and ridges are created and destroyed. ...Bruce Heezen synthesized the new data in relation to the expansion hypothesis in 1960, and I [Menard] did the same for the sequential hypothesis.[8]

The second- and third-last sentences of this excerpt show Menard's resignation to a fate of not having resolved all the issues to his own satisfaction. It is, in effect, a statement of capitulation. When all three proposed hypotheses are accepted as being sometimes true, without specific reference to data and differentials, then the process of active critical investigation has been abandoned.

Bruce Heezen's hypothesis of Earth Expansion appears to have contributed considerably to the ferment of the Plate Tectonics Revolution. Elsewhere Menard has characterized Heezen's synthesis regarding Earth expansion as somewhat indecisive. He bemoans the fact that Heezen has presented papers favoring expansion while also offering alternate hypotheses, such as classical continental drift and mantle convection.

How much of Bruce Heezen's conflict with Maurice Ewing, at Lamont, has been the result of rebellious temperament? And how much of that temperament was a result of having been diminished for preferring at times the Expansion hypothesis? Or, how much of Heezen's vacillation between theories was inflicted on him by his mentor, in the form of research assignments? The answer to these questions will probably never be fully known. The last joint paper published by Heezen and Ewing, in 1961, after the publication of which Heezen refused to co-author with his former mentor, discloses their difference:

---

[8] H. W. Menard, *The Ocean of Truth...*, Princeton, 1986, p. 132.

Ewing favored "a mechanism drawn by mantle convection currents," while Heezen believed "that the extension results primarily from the internal expansion of the earth."[9] This last joint paper appears to mark the moment, at the Lamont Observatory, at which the Expansion hypothesis became an unacceptable assumption. A lot of ego and personal pride was at stake.

Maurice Ewing's hope was to find ocean floors older than Jurassic. He made no secret of what he was after. In 1963 he described his own agenda approximately as follows:

Some of us hope to find a record of much earlier times, believing that the rough surface of the solid basement rocks beneath the deep-sea sediments may be billions of years old and, in fact, may be the original surface of the planet.[10]

While ocean floors, that are billions of years old, have still not been found, a small stretch of deep ocean floor in the eastern Mediterranean, dating back to the Triassic, has now indeed been certified on NOAA maps. But this limited occurence of something slightly older than 200 million years does not change the overall story of the oceans by much. It certainly falls short of the billions-of-years-old floors that Maurice Ewing was confident he would find. But the ocean floor that would be "billions of years old" was never found. Already in 1963 that same scientist, Ewing, found "thin sediment on bare rock ridge crests, suggesting spreading of some sort. He had not found any disturbance of sediment in the trenches, suggesting… that it was not being pushed into the continent." **Menard therefore commented, with some glee, that in those days Maurice Ewing "seemed to be proving that the earth was expanding."**[11]

---

[9] *Ibid*, pp. 106-107. Reference is made to Heezen, B. C. and M. Ewing, 1961, p. 640. "The mid-ocean ridge and its extension through the Arctic Basin," in *Geology of the Arctic*, Toronto: University of Toronto Press.

[10] Ewing, M. "Sediments of Ocean Basins." In *Man, Science, Learning and Education*, 41-59, 1963; quoted in Menard, *The Ocean of Truth…*, p. 201.

[11] Menard, *The Ocean of Truth*, p. 203.

The fluid state of Plate Tectonics theory, in those early days, was quite evident. Even the famous Tuzo Wilson, at one point, covered his bases for the eventuality that some day he might need to go the "expansion" route. According to Menard, he proposed "expansion of the earth along mid-ocean ridges, but on an acceptable scale."[12] It was the timescale that was objectionable in those days, not so much the process of expansion itself! Finally, at the very conference where the Eltanin-19 results were announced, Fred Vine referred to convection cells as being "presumed" and "mythical."[13] We are left to contemplate the implications of such statements for as long as current versions of Plate Tectonics are being taught in our schools.

In any case, there is ample evidence that, at the moment of the Eltanin-19 victory, in 1966, when Heirtzler and Pitman discovered parallel magnetic anomalies and began to establish sequences of ocean floor spreading, nothing beyond ocean floor spreading itself was theoretically fixed.[14] The speed of spreading was still debated, and at least for Vine and Heezen, convection cells and subduction processes seemed less likely.

The exorbitant rate of Earth expansion that the young oceans called for, to the effect that all the deep ocean basins were added to the planet in the course of a short 200 million years, was a notion that lay quite outside the reach of the Plate Tectonics revolutionaries. Even Heezen, as Menard bemoans that fact, never formulated his expansion theory in a suffici-

---

[12] *Ibid.*, pp. 147-151.

[13] *Ibid.*, p. 276.

[14] Eltanin-19 refers to the nineteenth research excursion of the ship Eltanin, cruising under the auspices of the Lamont Observatory of which Maurice Ewing was the director. They returned with excellent records of symmetrically-mirrored magnetic reversals along mid-ocean spreading rifts.

ently clear and scientific manner. He has not added anything new to what S. Warren Carey had already published.[15]

The Eltanin-19 discovery generated enormous enthusiasm. And when the excitement of this breakthrough rolled over the participating scientists, certain considerations, which until then had been important checks and balances, suddenly ceased to be of great concern any longer.

Among the first issues that were disregarded was the expansion hypothesis of Heezen. It was most easily rejected because of the personal estrangement that is said to have developed between him and the director of the Observatory. Another issue was the ephemeral or transitory character of the spreading ridges in the Pacific that all along had impressed Menard.[16] And third, there was the general youthfulness of the ocean floors, together with the tranquility that characterized sediments in the deep trenches—facts that even Maurice Ewing had admitted, in 1963, while he also was still hoping to find ocean floors that are billions of years old. His hope transcended the facts.

None of these crucial data seemed to matter anymore in light of the new knowledge and excitement about the fact of actual ocean floor spreading. "Create ocean floor crust now and worry about the precise manner of its disposal later," appears to have been the unspoken attitude. It all paralleled nicely the manufacture of atomic bombs in those days, with postponement of the problem of waste disposal. Perhaps the two mindsets—agreeing that a problem of disposal should never stop a

---

[15] Menard, *The Ocean of Truth...*, pp. 149-150.

[16] Menard's long insistence on the ephemeral character of the mid-ocean ridges, based on his Pacific data, is most interesting from the point of view of my hypothesis of Expansion Tectonics and Pacific evacuation. The patterns of spreading ridges in the East Pacific, and of ridges that surround Antarctica, are indeed younger than overall in the Atlantic. The great Eocene tectonic upheaval has, in the Eastern Pacific and Southern Ocean created conditions for fresh crusts and spreading ridges to form.

process—were not entirely unrelated. But then, the euphoria of winning cannot last, and neither can it obscure forever the possibilities of finding unexpected data.

It was for human-social-scientific reasons, ambitions and emotions that "convection" currents in the mantle and the "subduction" of ocean floors got accepted and became primary theory. And who can blame these scientists? No law of physics is known that can render possible an increase in the planet's mass and volume at a speed that satisfied their new ocean floor chronology.[17] Who in his sane political mind would risk his academic reputation by espousing a theory that disregards the revered truths of school physics? Physics reigns as uncontested queen among the natural sciences. It has done a superb job identifying forces that can be harnessed for the modification and destruction of the Planet; but it has remained amazingly unable to understand, or yield to, the natural forces of creation which also are at work.

The philosophical question of mechanism, vis-à-vis dynamism, continues to lurk in the background of this debate. In this regard the Earth Expansion hypothesis stands at least as strong as—or at least not any weaker than—the presently popular hypothesis to which everyone since Eltanin-19 has been flocking. "Convection currents in the Earth's mantle" are a mechanical metaphor that is applied ambiguously, apparently to conceal the fact that popular Plate Tectonics also has no known

---

[17]This is not to say that theoretically the possibility for an increase in mass and volume cannot occur. The simplest notion implies nuclear and chemical reactions happening in the mantle, which are feeding on older more compact mantle materials or possibly on the core itself, thus creating a more volumi-nous mass. But then there is also the presence of neutrinos and other micro-particles, penetrating the Planet from the greater universe. They do offer the possibility that mass or volume be added for saturation, perhaps as the result of their entanglement in the viscous environment of the upper mantle.

dynamic. The mechanically slanted metaphor of "convection currents" simply conceals ignorance regarding the real dynamic that might empower magma formation and movement. Earth expansion and convection currents in the mantle are both mechanistic concepts. In either case are the dynamics which are implied unknown.[18]

The demand that any new geological theory should offer a full disclosure of its implied dynamic sometimes does come across as being disingenuous. First, there is the typical ambiguity that is being maintained between dynamics and mechanisms. The general word "current" implies some kind of movement merely by ambiguities inherent in the word. The hypothesis of convection currents therefore merely begs the question, of whether a fictitious hypothetical process is alluded to here in order to rationalize a postulated movement.

The new global seismic tomography shows areas of greater viscosity or hardness in the mantle. [19] These areas are found

---

[18] Copernicus and Kepler described the mechanism of our solar system, and for five centuries we have evolved astronomy from their discoveries without being able to understand the dynamics. Of course, since Newton became fascinated with gravity we have increasingly been talking about the dynamic dimension. But mathematical ratios are not the dynamic itself; these are simply "mechanisms" constructed of numerically imaginable and mentally manageable fantasy "handles." A serious ethnological study of the evolution and the uses of mathematics, if it were conducted, would reveal that the application of numbers to nature has always been a theoretical handle for gaining control—and not for acquiring neutral understanding. You *name* your potential equal beings and you *count* your possessions. Let anyone provide a precise and provable explanation of our planet's dynamics, of the "force of gravity" perhaps, and I will need to rewrite this note, with pleasure.

[19]Compare Stephen P. Grand, Rob C. van der Hilst, and Sri Widiyantoro, "Global Seismic Tomography: A Snapshot of Convection in the Earth," in GSA Today, April 1997; and Rob C. van der Hilst, Sri Widiyantoro and E. R. Engdahl, "Evidence for Deep Mantle Circulation from Global Tomography,"

at the wrong places for helping establish "convection currents," "deep mantle circulation" or "ocean floor subduction."

Unfortunately, two conflicting unproven hypotheses do
not add up to a proven theory. Thus, by working on the problem
of Earth expansion we are still functioning in the realm of hypotheses. Nevertheless, trusting the scientific mind as to some extent
I still do, I believe that it will be only a matter of time before a
theory of "ocean floor spreading without subduction" will claim
its data base in the pursuit of general Spread and Growth Tectonics. I suspect that some of the loosely seated freeloader hypotheses will sooner or later be shaken off and fall away.

The new ocean floor chronology has been available since
1985 from such map compilers as R. L. Larson, W. C. Pitman
(III), X. Golovchenko, S. C. Cande, J. F. Dewey, W. F. Haxby,
and LaBrecque, in *The Bedrock Geology of the World*, Freeman and
Co., New York. Internationally, the information was made available, in 1988, with the *UNESCO Geological World Atlas*. And
finally, in November 1996, the NOAA map "Age of the Ocean
Floor" was distributed at the Denver meeting of the Geological
Society of America. This author was present.

The realization, that there was an Eocene Tectonic Event,
dawned on me while contemplating the isochrones on my chronological globe—especially regarding the Pacific, the Indic, and
the Southern Hemisphere. Separate special chapters will feature
significant data for each of these oceanic regions. The Bight of
Australia still faces the cape of its former partner, South America
(Figure 6). This layout, admittedly, was made complicated by
the inverted intrusion of Antarctica.

---

in *Nature*, vol. 386, 10 April 1997. It should be said that even if some type of
deep mantle circulation exists, it is not necessarily the kind that could
prove ocean floor subduction.

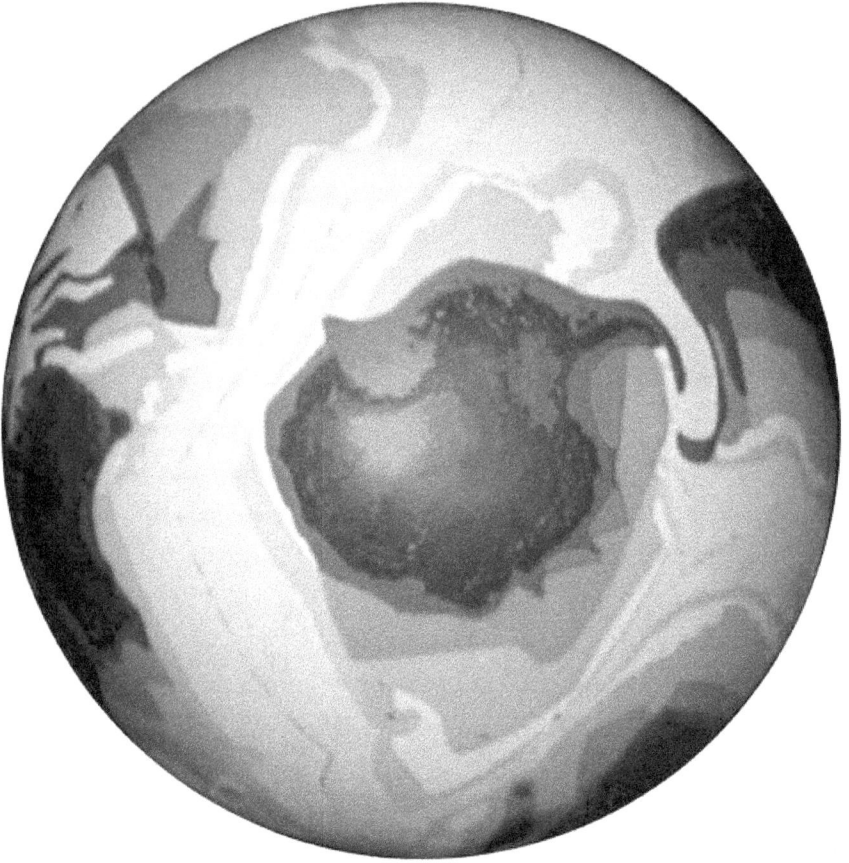

Fig. 6. Isochrone globe, showing the southern hemisphere with the Antarctica "Plate." On this globe, painted by the author, the Paleocene is indicated in a lighter tone than the Eocene. The derivation of Upper Cretaceous seafloor, at the western back of Antarctica, from across a young spreading rift (see Figure 29), appears obvious from this view.

3

From Ocean Floor Chronology
to Expansion Tectonics

When in 1979 I published my first essay on the expansion of Planet Earth, I was still unaware that somewhere an ocean floor chronology was being assembled. I based my work entirely on common sense geology and on matching continental contours.[20] At that time I was even unaware of the fact that others before me had attempted to make terrella models. Soon it became obvious that I was not the first creature on Planet Earth who independently had noticed evidence of its expansion. If I were to do historical research regarding this question, I reckon that I could identify several dozen of us. Each time, since 1979, when I heard of yet another independent discoverer, I was elated. At least humanity's judgment regarding my mental state was not unanimous. But my conclusions have turned out to be somewhat different than those of my co-discoverers. **Today there may be three people who agree that the new ocean floor chronology holds the best key for demonstrating ocean floor spreading and Earth expansion as well.**

I found the UNESCO Geological World Atlas with its ocean floor chronology about five days after the Northridge Earthquake of California, during January of 1994. At that time, as on many other occasions, I was hoping to understand experts

---

[20] Karl W. Luckert, *Mother Earth Once Was a Girl: a Scientific Theory on the Expansion of Planet Earth.* ATR, Supplement 1. Flagstaff: The Museum of Northern Arizona Press, February 1979.

who publicly explained this geologic event. I began working with the new ocean floor chronology immediately. To this day I feel embarrassed about the fact that this information was resting in our university library nearly five years before I found it—and to my astonishment, I was still the first person to break open the pages and the maps.

Shortly after my exposure to the isochrone maps of this Geological World Atlas, in 1994, I wrote an essay to summarize for myself the very first lesson I learned from the new ocean-floor chronology. It was published in 1996 by Theophrastus Publications in Athens.[21] The realization dawned on me rather quickly that there must have occurred something like an Eocene event; though, in this first essay I still referred to it as "Pacific Evacuation"—meaning, of course, the evacuation of Antarctica. Figure 7, here, represents a compilation of three figures of this 1996 essay. Presupposing my path of inquiry, the isochrone data confirmed my Earth Expansion hypothesis quite adequately. I carried my message to the GSA (Geological Society of America) Annual Meeting in Denver, in 1998. In subsequent years, I displayed exhibits and distributed pamphlets at all five Regional Conferences of the GSA. At the 1999 GSA Annual Meeting I distributed free copies of the treatise that served as prototype of the present book, which I published in that year—*Planet Earth Expanding and the Eocene Tectonic Event*.[22]

Now, in 2016, there are still four points on which I find myself standing alone, for which I must take personal responsibility, and without these four points I would probably remain unconvinced of Earth expansion.

---

[21] Karl W. Luckert "A Unified Theory of Earth Expansion, Pacific Evacuation and Orogenesis," in *Theophrastus' Contributions to Advanced Studies in Geology*, pp. 61-73. Athens, Greece: Theophrastus Publications, S.A., 1996.

[22]*Planet Earth Expanding and the Eocene Tectonic Event*. A Lufa and Triplehood Publication, 1999. Also posted at www.triplehood.com.

Figure 1

The Earth, approximately 135 Mya, showing only Jurassic Oceans

Figure 2

The Earth, approximately 70 Mya, showing a north-south spreading rift divide the Pacific

Figure 3

The Earth at present, showing Jurassic, Cretaceous, and post-partition ocean floors. A Paleocene divider runs midway down the Pacific

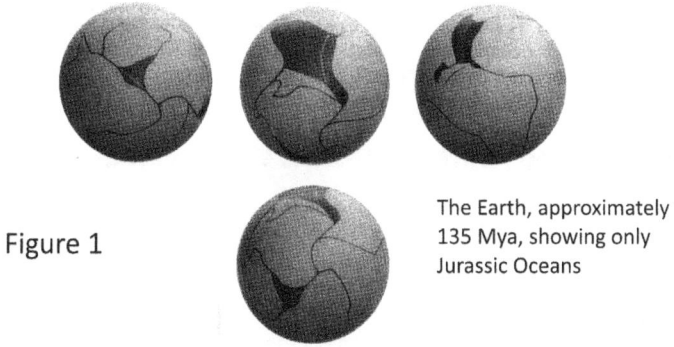

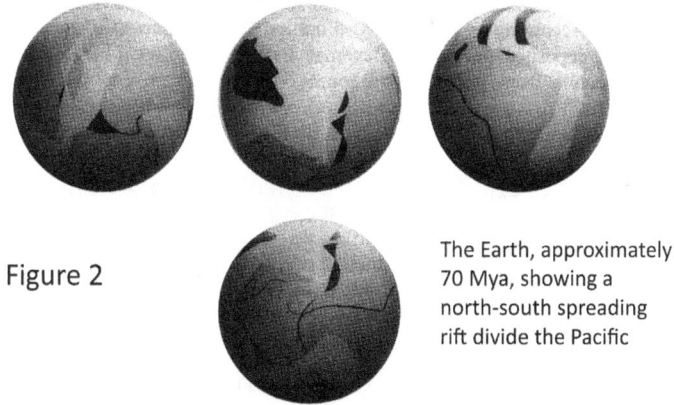

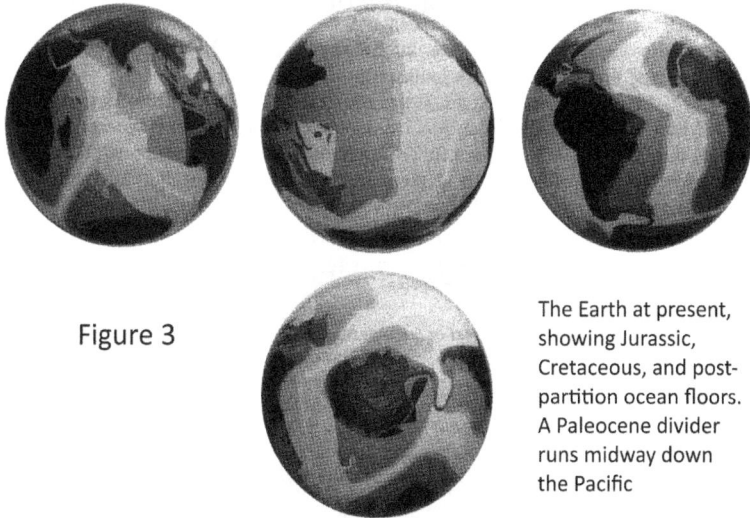

Fig. 7. Earth Expansion in three stages. Page 68 of the author's essay, in Theophrastus Publications, 1996.

Point One: The Pacific opened from West to East. Antarctica has twisted itself out from the expanding Pacific space. It was a roundish segment of the original continental crust. **Isochrones provide data for the ocean floor chronology, that is, for the sequence by which the Pacific Ocean was opened.**

Point Two: Australia sat at the cape of South America. This is a view I have held since 1979. The severance of these two continents from each other has triggered a sequence of position adjustments involving Antarctica, Southeast Asia, Australia, South America and even North America. By contemplating the chronological maps, which the Plate Tectonics Revolution has produced—and which most heirs of that revolution seem to be ignoring—I have discovered that their isochrones do in fact support the Eocene Event quite nicely, on a global scale.

Whatever explanations of Eocene ocean floors one might point to, they are not mere topics for rhetorical consideration; they are challenges for evolutionary scientific inquiries. They should either be verified by scientific exploration or be rejected thereby. My overall delineation of the Eocene Tectonic Event, based on isochrones, presupposes numerous points of former continental connections. I recommend that all these be subjected to thorough geological examination and comparison.

Point Three: Continents are torn apart, not born apart. I have become convinced that for continental separations to have occurred, upon an expanding planet, a large field of horizontal stresses needed to be resolved. The tearing out of Antarctica from the original Earth crust, and severances of any continents from others, must be accounted for not only on the accustomed terms of matching continental areas and contours, but also within the total budget of a gradually decreasing intercontinental cohesion. Factors of elastic cohesion by pre-magmatic viscosity, at various higher and lower pressures in the upper mantle, may be mentioned as belonging to that budget. The tensions that were relieved by past overstretching, by rifting and by tearing, may be subtracted from this budget. Continental segments that were broken away from the planet's total area of original crust may also be subtracted from that budget of cohesion stresses. **The continents were torn apart. They were not born apart to drift in conformity with Wegener's *Vagabund* paradigm as "wandering continents."**

Point Four: Deep trenches result when vaulted ocean-floors collapse. Where in association with smaller transverse faults, or spreading faults, a mid-ocean spreading rift cannot relieve horizontal expansion stresses, along the irregular or curved boundaries of an ocean floor "plate," and **if an ocean plate finds itself sitting on the flattening mantle of the expanding planet, with an overly acute arch, then the plate is destined to collapse. The collapse may crack open trenches along the peripheries of ocean floors, splitting from the surface downward, possibly even through the entire thickness of the crust.** However, the trench will immediately be cooled by ocean water, likewise from top downward. At the underside of the fracture, along the upper asthenospheric boundary, a patch of thickening rock-melt will likely accumulate to seal the wound.

*   *   *

Henry W. Menard, in 1986, erupted in writing jubilation history. He indicated that at some point "the revolution was over; the flowering of geology could begin."[23]

From this author's perspective the Plate Tectonics Revolution appears to be only half finished. But, let Earth Science blossom all the same! And would, that the smoother pebble which this lonely wanderer along the shore of that same great ocean of truth, has noticed within the heap of data that celebrating

---

[23] H. W. Menard. *The Ocean of Truth...*, p. 294.

revolutionaries have neglected, may also unfold as a flower and ripen into a fruit, in the field of scientific labors!

From among all the people who profess a theory of Earth Expansion, as far as I know, my approach comes closest to that of James Maxlow, in Australia. Both of us assign basic importance to the new ocean floor chronology; both have worked with the same chronological data and with similar maps. Nevertheless, our respective expansion sequences have turned out differently. Inasmuch as we have arrived at our respective conclusions quite unaware of one another, this happenstance may turn out to be an advantage for both in the end. In science it is equally important to learn what something is not, as it is to explain what something is. Differences among our conclusions can stimulate adjustments in the future. Without objections raised by others, this booklet, and even this chapter, would surely have been written less well.

Our differences of opinion regarding the Pacific-Antarctic and the Indic oceans are considerable. I concede that Maxlow's solution for the northern Pacific would be an improvement over Klaus Vogel's hypothetical slip-fault between the Asian and North American crustal masses—that is, it would be an improvement if it worked. Following Wegener, **both Maxlow and Vogel match the Bight of Australia with the round of Antarctica and then place Australia into the North Pacific.**[24]

The placement of Australia, on Maxlow's Jurassic globe, as well as the manner of emancipation of Southeast Asia from the Asian mainland, versus my own insistence on a worldwide

---

[24]A comparative DVD video, on Earth-Expansion hypotheses, by Karl W. Luckert, titled "Four Theories of Earth Expansion and the Eocene Event," was distributed at the Urbino (Italy) Conference in 2004, "New Concepts in Global Tectonics." The Video Script is downloadable at <www.triplehood.com>.

pattern of continental cohesion constitute, probably, our greatest differences of view. According to Maxlow's terrella sequences (January 1996, especially pages 43 and 50), the opening of the Jurassic Pacific derives Australia from alongside North America.[25] All the while, his southbound Australia pulls a pre-loosened portion of East Asia southeast-ward, to create present-day Southeast Asian shapes. This arrangement would entail pulling Borneo and the Philippine Islands away from Japan, and Sumatra away from the coast of China.

I can appreciate this line of reasoning, and how the puzzle of the Northwest Pacific would possibly recommend such a solution. For a while, in 1994, I myself considered twisting New Guinea out from the Philippine Sea. But that would have run against the logic of the ocean floor chronology in that region. Had Australia pulled all of Southeast Asia out of the hide of East Asia, I reckon, the northward-squeezed Mariana Scar would never have come into being and the Java Trench would now likely curve in the opposite direction.

Instead, I explain the overall bent shape of Southeast Asia and the existence of the "marginal seas" along the East Asian coasts, as the result of a far-reaching Eocene tectonic event. Eocene ocean floors, on Pacific and Antarctic hemispherical maps tell the story of this event. I derive the bend of the Sunda Shelf from along the straight Ninety East Ridge in the west—bending and sliding away from, and clearing, the triangle of Eocene ocean floor.

The western edge of Southeast Asia, according to my view, calls for chronological rethinking (see Figures 15, 21a). **I surmise that Sumatra, Java, and even the western edge of Australia, were formerly aligned and stretched southward**

---

[25] James Maxlow, *Global Expansion Tectonics: Small Earth Modelling of an Exponentially Expanding Earth*. Glen Forrest, Australia: Terrella Consultants, January 1996. See also Maxlow's new (2015) website: www.expansiontectonics.com. His interpretation in 2015 appears to be essentially the same.

along the present Ninety East Ridge. The latter extended southward and apparently included the so-called "Broken Ridge."

The triangle of ocean floor, bounded by Sumatra and by the northern third of the Ninety East Ridge, happens to be Eocene. This means that Southeast Asia, prior to the Eocene epoch, was stretched southward along the line of that Ridge for a considerable distance. Southeast Asia must have gotten bent eastward at the very moment when the central spreading rift of the Indian Ocean began to change to its new diagonal alignment, inclining from Southeast to Northwest. This realignment, too, was forced by the primary active continent during the Eocene Event —by the intrusion of an ocean floor plate which, attached to Antarctica, came into the Indic.[26] At the same time, during the Eocene, Asia's marginal seas were spread open some more.

Also, the oldest fragments of ocean floor in the eastern Pacific, along the American coasts, are Eocene. (Figure 21b) The eastern Pacific became a new geological entity. All of this suggests that during the Eocene a major tectonic event occurred.

But where could Australia have come from if it did not come from the northern Pacific? My own answer was given already in 1979, based on the consideration of continental contours. It was my luck, back then, not to have been influenced by Wegener's work. The Bight of Australia still faces its partner, the tip of South America. Of course, this layout has been made complicated by the intrusion of Antarctica. According to my hypothesis, **a longitudinal Earth-encircling resilient "belt of continents" comprised of the Americas, the Arctic, Asia and Southeast and Austral Asia was broken, along the Bight of Australia and the Cape of South America.**

---

[26]The Antarctic Plate consists of two parts—the roundish continent itself, and a triangular accretion of ocean floor that was added since the Lower Cretaceous while the round continent was getting peeled away from Alaska.

At the opposite side on the globe from South America, Australia and Southeast Asia rebounded north together and veered east toward the stretch that was vacated by the departing Antarctic Plate, in the eastern Pacific. At some point in time, around 42.7 Mya perhaps, the Antarctic Plate twisted counterclockwise and southward.

For a roundish piece of crust, sitting on an expanding sphere, slipping while twisting may come easier than straight forward bulldozing of an entire continental width. Thereby Antarctica backed its tail (e.g. heel) past Cape Horn as far as the Sandwich Islands, some 2800 kilometers into the South Atlantic. There was no resistance to this intrusion. A hundred Ma earlier this area was weakened by the departure of the African continent. Antarctica has since returned to a distance of about 1000 km. Meanwhile, the northern tip of Antarctica's accreted ocean floor plate came to rest in the southern Indic Ocean, pointing north again and leaning toward the south-eastern shores of Africa.

The exact inclination of Australia/New Guinea, as "buckle" of the longitudinal global belt, prior to the Eocene event, has not yet been firmly determined. Presumably the Ninety East Ridge, the Broken Ridge, and a stub of ocean ridge at Australia's southwest corner, will need to be re-matched. The outcome of this unfinished task will depend on more precise ocean floor topography in the Broken Ridge and the Celebes Swirl areas—as well as on what the untwisted landforms of Southeast Asia in general would have looked like in their original positions.

In addition to the original location of Australia, in relation to South America and Antarctica, there are implied between Maxlow's expansion hypothesis and my own, different theoretical "budgetary" assumptions. These pertain to crustal tensions and cohesions that must be accorded to all original crusts on this expanding sphere. Aside from spatial considerations, and the geometry of contours, the Earth expansion process implies a budget of tensile forces that must be kept in mind for all continental severances. **Various shapes of continents were torn from the Planet's crust under different and changing tensions—that is, among adjacent tensions which themselves were being constantly altered by Earth expansion and by localized tearing. Continents were not born apart simply to drift or to associate wherever on a globe one could make them fit, or aimlessly to wander about in Panthalassa. Continents were torn apart!**

Continental shapes certainly did not come about to fit some Platonic norms for ocean contours, but rather, ocean water flowed into spaces wherever continental crust was being torn by tension. **The process of tearing free continental slabs—upon a resilient asthenospheric surface—required forces and implied directions which for every tearing-event needs to be explained structurally, in agreement with the isochrone sequences that, meanwhile, have gotten established by the science of Plate Tectonics.**

Because the circumference of Planet Earth was increasing, substantial lines of fissure were forming along the bottom of the crust. Such crevices formed wherever, as the result of Earth-expansion, a domed portion of the crust needed to sink, to flatten and to adjust to changing curvature. Such adjustments frequently required faulting or downward folding. When later there were deeper oceans with younger and thinner floors, they too came to rest upon squeezed magmas and creeping rock materials. Flow-materials, such as lavas, broke through the thinner ocean floors much more easily than they could break through thicker continental crust.

The Geologic Time Scale and Periods of Extinction

Briefly I must mention the possibility that the geological timescale we are using, ranging from old Jurassic seafloors to the exit of the Antarctic plate during the Eocene, may at this point in time only provide an approximation. To improve pale-ontological sequencing, our "Eocene Transition" hypothesis will someday itself need to become engaged in the review process. The periodization of "isochrones," where it now stands, has been accomplished by geologists in cooperation with paleontology and it remains an art that is still evolving. It is based on what so far has been learned about a sequence of universal extinctions of animate beings, by periods and epochs. Ocean floor cartogra-phers assign their isochrone data to periods and epochs that have been pre-identified by paleontology. Working toward a universally valid periodization, one finds that the evolution of life forms indeed appears punctuated by discontinuities and extinctions. But then, while ascribing time intervals from iso-chrone readings, it may occasionally, for questionable timespans be better to substitute blank spaces or question marks.

It is possible that ocean floors, which episodically have been disturbed by massive movements of crust, could intermit-tently have left the environment in a hostile state toward various forms of life. We do not really know how clean or how friendly toward life the natural processes have kept the oceans. After the Eocene transit of the Antarctic Plate, southward along the eastern Pacific, there may exist a good possibility that the general climate and living conditions on Earth have been affected, or that a food chain would have been rendered dysfunctional.

And there is another consideration. It might have taken some time for the youngest ocean floors to harden, for lavas of a spreading rift to embed additional magnetic reversals. We do not know how quickly continental separations have happened or how quickly a large departure wound could heal after it had been torn. At least in the southeastern Pacific, an area across which, some 42.7 million years ago a large continental plate has scraped and bared the asthenosphere, our people on drill ships continue to struggle to find coherent patterns of data.

Interpolation into data-less areas is not always safe. There are reasons to be cautious about what a spreading rift and its adjacent magnetized isochrone stripes can actually tell us. After a continent has slid over it, how long might it have taken for the flank of a rift to harden and to begin registering magnetic reversals again? A prolonged soft period could easily skip a round or more of those magnetic reversals. It is conceivable that there are paleontological as well as magnetic gaps in our time-scale. Unknowns in science are treated as though they never were. On the other hand, recognized gaps in the data do recommend that all steps of time-sequencing should be kept under constant review.

There exists, of course, also the problem of alignment for the magnetic directions in the crusts themselves. What happened when the Antarctic Plate twisted away from the coast of Alaska, down into the southern Indic? Do magnetic reversals go undetected when 45, 90, or 135 degree twists occur? Of course, we know that it is not just the reversals, but mirrored patterns of reversals that count. When the Eocene Tectonic Transition is studied geologically in earnest, regarding any data from this period of upheaval, a second look at this question will be due.

It is reasonable to suspect that various upheavals on the ocean floor crust, during the past 200 Ma, could have caused multiple localized episodes of extinction. Our general ignorance concerning the destruction of environments moves us to hasty analogical reasoning. For example, the extinction of animate life that, supposedly, a single meteorite in Yucatan (the Chicxulub Meteorite) has wrought, would probably pale in comparison with changes that were associated with the Eocene Transition—including the simultaneous opening to the world ocean. The huge global belt of continents, breaking in the south, united the oceans of the planet into one. Presumably it opened the world ocean for colder water circulation. The Eocene event moved four continents by hundreds and thousands of kilometers. According to isochrone maps, it happened in the middle-to-late Eocene. The maps cannot tell us whether the magnetic poles themselves were moved—or whether they even could be moved.

GEOLOGIC TIME SCALE

EON	ERA	PERIOD			EPOCH	Age in millions of years before present
Phanerozoic	Cenozoic	Quaternary			Holocene	Present
						0.01
					Pleistocene	
						1.6
		Tertiary	Neogene		Pliocene	
						5.3
					Miocene	
						23.7
			Paleogene		Oligocene	
						36.6
					Eocene	
						57.8
					Paleocene	
						66.4
	Mesozoic	Cretaceous				
						144
		Jurassic				
						208
		Triassic				
						245
	Paleozoic	Permian				
						286
		Carboniferous	Pennsylvanian			
						320
			Mississippian			
						360
		Devonian				
						408
		Silurian				
						438
		Ordovician				
						505
		Cambrian				
						570
Precambrian		Proterozoic				
						2500
		Archean				
						3800
		Hadean				
						4550

Fig. 8. Excerpt from GSA Geologic Timescale: http://
geomaps.wr.usgs.gov/parks/gtime/index.html

Paleontology has been working toward a globally valid time-scale. But perhaps for different oceans or inland seas, some localized time scales should be attempted first. For example, animals that lived in the Mediterranean or Black Sea area were less affected by the Eocene Transition, or at least affected later, than those that lived in or near the eastern Pacific.

Plate Tectonics and Earth Expansion Perspectives

Academic arguments, controversy, and distortions, about scientific considerations concerning the Earth's stable size, versus its enlargement in size, are in full swing. In the passion of confrontations, many hypotheses are being upgraded to theories far too quickly, at both sides in this debate. Suggestions are being accepted or rejected for organizational, professional or existential status, rather than for seeking truth about the geological subject matter at hand. Hypotheses are seldom being collapsed far enough, down to basic data, for thorough verification.

In the study of geology, along the Plate Tectonics front, a strong founder complex is in place. The name of Alfred Wegener shines up front and ranks high. By contrast, among Earth Expansion hypothesizers the need for a single founder figure has not yet arisen—even though Donald R. Prothero, at the Plate Tectonics lineup, has identified the once prominent Australian geologist, S. Warren Carey, as the last legitimate scientist to be taken seriously on the idea of the expanding Earth.[27] This style of limiting the forum is obviously intended to benefit the disparagers of the Earth Expansion idea. It seriously limits the hypotheses which are permitted to be brought into the discussion. Fortunately, this author is not affected by these constraints, because for the length of this treatise he is restricting himself voluntarily to the database of the better organized Plate Tectonics science.

———————————————

[27]"Donald Prothero, "Cracking earth and crackpot ideas." *Skeptic Magazine*, vol. 18, number 1, 2013. www.donaldprothero.com/files/92370160.pdf.

Over the years I have personally gained the impression that more of the Plate Tectonics data do support my Earth Expansion hypothesis, on more points than the Ocean Floor Subductionists stand to profit from their own mapping labors. This is not the time for us to go back a hundred years or so, to the manner in which a certain founder knew or imagined the facts. The Plate Tectonics Revolution, meanwhile, has gathered a database of its own, precise enough to enable a fresh start. Of course, both sides in this debate fail to explain the first axioms of their positions.

No Earth expansionist knows for sure why the Earth would expand. And no ocean floor subductionist knows how, or even where, in the course of the past 300 Ma, ocean floors on this planet could have been subducted. Nor can Wegenerians explain why a natural process might have shepherded all wandering or drifting pieces of a jig-saw puzzle to one side of the globe.

It is also not known why Plate Tectonics advocates, by and large, are neglecting their own self-created ocean floor chronology. Since the days of Heezen and Carey, how many among them have even tried to disprove a supposedly "false Earth Expansion hypothesis" in light of their own new isochrone maps?

My own approach to the question of Earth expansion was from the outset driven by a search for physical evidence. My initial interest in building terrella models subsided rather quickly until, in 1994 I found the isochrone maps which Plate Tectonics scientists have been producing. I examined the chronological data and became convinced that Earth expansion must either be proven or disproven by these data. The isochrones provide rather precise information about ocean floor sequences and expansion.

If henceforth a zero-ocean terrella model is to be constructed, then its production must happen by way of subtracting isochrone stripes from ocean floor areas in accordance with a reverse sequence of the magnetic time scale.

Some "Earth Expansion" hypothesizers maintain a more cautious skepticism toward the new chronological maps of Plate Tectonics than I do. Carl Strutinski (2015), for example, rightly points to the ambiguity of the Pacific isochrone stripes. On that account he tends to doubt the general usefulness of isochrone maps for theorizing about Earth expansion.[28] His specific identification of the Eastern Pacific seafloor as a problem area, is undoubtedly correct. I myself ran up against this problem in 1994 and I suggested my sequential explanation to it, in my 1999 publication, by way of adding question marks at two of the most problematic stripes (reproduced also herein, as Figure 21a-b). However, only because a half-ocean map turned out to be incoherent does not mean that all isochrone data should be avoided in the remaining parts of the world ocean.

My personal solution to this East-Pacific problem is not to disregard isochrone maps in general, but from the point of view of my "Eocene Transition" hypothesis to resolve, sequentially, how this problematic magnetic non-pattern could have gotten into the ocean floor in the first place. From the perspective of my hypothesis a simple and rational explanation is possible. Had the drill ships and the map makers thought of this possibility, I am convinced they would have left a few stripes on their East Pacific Map uncolored, as *ignota spatium temporis,* and moved on with the remainder. We are confident that the map makers will catch up with clarifying this matter sooner or later, by way of drilling key spots which are implied by our enhanced approach to expansion tectonics.

––––––––––––––––––––

[28] Carl Strutinski, "Some Reflections on the Charts of the Ocean Floor: do they hide more than they reveal?" at <www.dinox.org-publications-Strutinski2015.pdf>

4

Continents and Mountains

A treatise on continents, on spread and growth tectonics, and on ocean floor chronology, cannot be meaningfully presented without providing some explanations about the formation of mountain ranges on land as well as of sea-mounts in the oceans. While it is true that Earth expansion happens for the most part beneath the crust, most evidence concerning the enlargement of the planet, including seismic magnitudes, must be observed and recorded near the surface of the crust.

Along the bottom of the Planet's original crust, oceans and mountains had nearly identical beginnings. They began as crevices at the underside of the crust. One type of crevice found space to widen upon the expanding mantle. Magma that went there has created water-cooled ocean floor. By contrast, constrained crevices under continental crust were filled with rock-melt and were raised as bulges and mountain ranges. Whereas mountain ranges on continents are cooled by climate and weather; sea floors, spreading ridges and seamounts are cooled, thickened and hardened, by ocean water.

Some Earth scientists begin their lectures about mountain formation by explaining adjustments in terms of isostatic vertical movements. This approach conveniently reduces three-dimensional questions to fit in two or even into a single dimension. Some teachers distinguish "isostatic" from "tectonic" movement —the latter is then explained as happening horizontally. Many students today take that to mean "two-dimensionally." Others explain mountain uplift more dynamically in terms of "thrusts" or "heaves." They think of tectonics and isostasy as a motor for continental movement and mountain uplift. Minds trained in general Plate Tectonics, shifting between the horizontal and the vertical dimension, do not necessarily produce inconsistencies. Inasmuch as the uplifting of mountains supposedly happens as upheaving with the angular subduction of ocean floors along Benioff Zones, the vertical and the horizontal dimension are averaged with the "lever-like" angular subduction.

All problems of the human mind reduce eventually to problems of language. But language cannot simply be invented or improved. It must grow by common understandings, against common misunderstandings, as well as by common consent. It took thousands of years before one could say "sunrise" and educated people understood that the Sun was not actually rising but that the Earth was rotating. After this knowledge had been printed in a book it took another few centuries until "rotation" was generally assumed. But today the Sun still rises, at "sunrise."

A year before this octogenarian author was born, Ott Christoph Hilgenberg wrote a book about the growing Earth. Today, to say that "the Earth is expanding" requires still more books to explain these words. The Earth has continents, oceans and mountains, and other distinguishable essentials. All basic knowledge about such matters must be readjusted, together, so that the full insight of "Planet Earth is expanding" can be understood and communicated rationally.

Folding, Flanging, Melt and Squeeze

In the past, three terms have proven useful to explain our brand of Spread and Growth Tectonics: 1) Tensile Folding, 2) Flanging, and 3) Episodic Expansion Squeeze (originally Relative Expansion Flow). With these concepts I explained, in my 1999 publication, some interactive aspects of Earth expansion— of ocean- as well as of mountain-formation.

Tensile Folding.—"Tensile Folding" is the first of three interrelated mechanical concepts of mountain formation about which I wrote in 1999. **Take a sheet of any fabric, stretch it evenly in the second dimension (i.e. lengthwise), and a pattern of parallel stretch-folds will appear. This is the starter model for obtaining the kind of parallel mountain ranges that rose, by and large, upon continental peripheries**—with the proviso, that continental peripheries, resting on an expanding and viscous asthenosphere, seldom stretch in straight lines and more often do so along curves and wrinkle-folds.

Tensile Folding does not lift up any mountain ranges by itself. It merely determines whether and where along the weakened undersides of synclines, under continental flanges, certain fissures might open, and where intrusions of flow-rock or magma might occur.

We can safely assume that some Earth expansion happened already in the Triassic, or earlier. During the earlier episodes, until the first deep oceans opened, it was simply the original Earth shell that was still somewhat capable of being stretched.

The process of erosion at the surface of the crust—where anticline folds tend to crack along their weakened crests and are worn there quicker by wind, rain, and ice—is tectonically similar to the vulnerability along the bottom of synclines. But under the crust all processes happen at higher pressure and temperature, inverted in space. Faults and fissures do form along the underside of crustal synclines and these, in turn, invite intrusions of rock-melt or magma from below. To continue with our inversion analogy, these faults and fissures invite a type of upside-down "erosion as well as sedimentation" of the kind that is possible in the pressurized, and transitionally decompressed, environment in the lower lithosphere and upper asthenosphere.

Circum-Pacific and circum-global tension has arranged the mountain ranges which run along the western flanks of the

Americas. Similar north-south tension has folded those that extend between Ethiopia and southern Africa. Tension has aligned the mountain ranges of Eurasia which, by and large, stretch east and west from the Himalayas to the Pyrenees, also those which run northeasterly from Southeast Asia to eastern Siberia.

If there was enough stress to accomplish Tensile Folding of continental crust, there might also have been other stretches of reinforced crust that could be pulled or stretched into submersion below sea level, or even stretched to a point of breaking. The same circum-Pacific tension, that has initiated the folds of the Rocky Mountains and of the Andes, also has elongated and pulled apart the continental shreds of Middle America. Likewise along the northern rim of the Pacific, between North America and Asia, the continental crust of the Arctic region has been stretched considerably. Southeast Asia was elongated southward until Australia was torn away from the tip of South America. Some of the elongation in the crust, underlain by viscous upper mantle materials, appears to have been elastic, while other spans were overstretched and deformed irreversibly.

Allow me to mention again the east-west tension between southern East Asia and the Pyrenees. The southern coastline of Asia is made complicated on account of three peninsulas or sub-continental flaps—India, Arabia, and Southeast Asia. Tensions that tore these three continental flaps have sent South Asia's latitudinal "tensile folding" further inland (see Figure 13).

Along this stretch, the Earth-shell was not sheared quite as uniformly as happened along the round Pacific Rim, nor to this day has it been completed. Africa remains attached to Eurasia along the halfhearted Mediterranean spread. Ever since Africa had lost its foothold at the toe of South America, early in the Lower Cretaceous, no anchorage remained for the large continent of Africa to pull away from Eurasia much farther.

Tensile folding, by itself, does not accomplish all there is to mountain formation. It accounts only for relatively low folds and initial cracks under synclines. Massive rock-flow and magma intrusions, plus extra quantities for general upward bulging, do require another process or mechanism that must be explained.

Flanging.—Our second concept of an interactive mechanism in the process of mountain formation is Flanging. Intrusions of magma into fissures and rifts, into the crust from underneath, and the subsequent rise of mountain ranges, participate in a process that interplays with Earth expansion as well as with the gravity of the planet. This mechanism is best understood as simply another aspect of Earth expansion. I call it "Flanging."

A continent is a fragment of the original smaller earth-shell, and flanging happens whenever a segment of continental crust finds itself situated on an expanding sphere, thus upon a growing and flattening substratum. While the sphere expands, the original curvature of continental crustal segments fits less and less on the decreasing curvature of the mantle.

As rim portions of the dome buckle and crack somewhere between the arched center and the continental perimeter, flange-synclines are getting folded at the underside of the crust.

Tensile folding under the perimeter enhances the crustal collapse as directional flow-patterns for Expansion Squeeze are getting established underneath. Some flow-rock will get diverted into crevices overhead or pasted under upward-bulging future mountain ranges.

For my first Earth Expansion essay, in 1979, I did several balloon-expansion experiments. I patted a round patch of oil putty onto a half-inflated balloon, gave it a coat of spray paint to simulate a crust, and then inflated the balloon some more. The outer edge of the batch of putty produced a flange. It curled upward and showed wrinkles along the top of the syncline (Figure 9). On a sticky asthenosphere this flange would have adhered to and decompressed the substratum underneath.

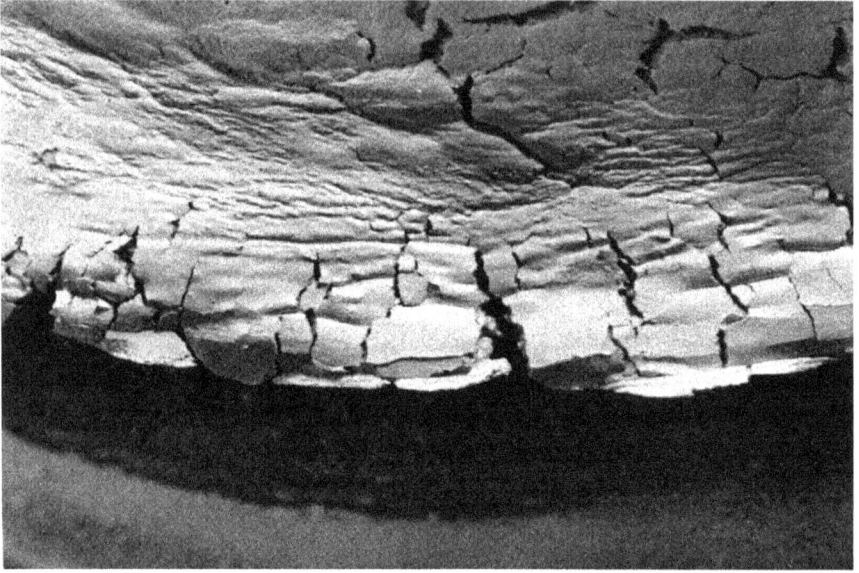

Fig. 9. Experiment 1979-a. Putty slab on an expanding balloon, showing top view of Flanging, with "tensile folding" in the syncline. In 2002 it became obvious that the cracks and folds at the underside of such flanges were more significant for future mountain building.

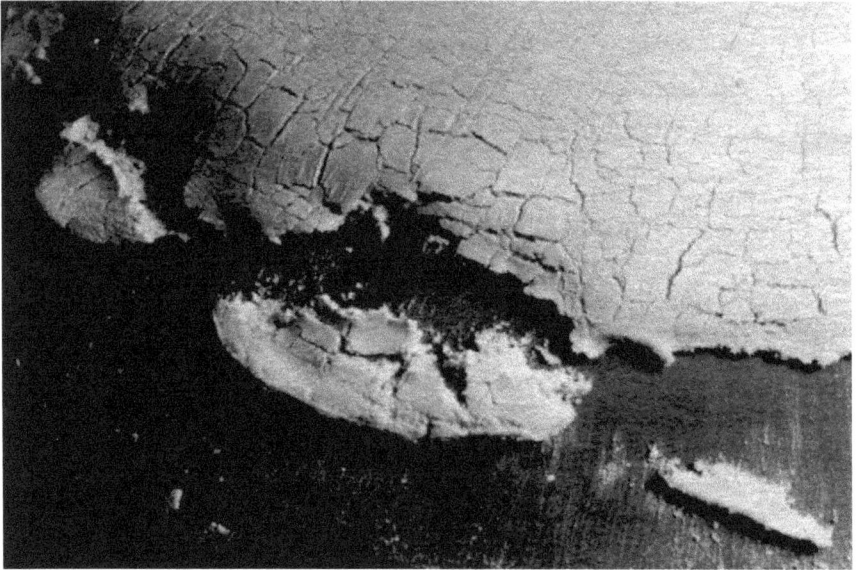

Fig. 10. Experiment 1979-b. Putty slab on an expanding balloon, held a little drier. Upon the expanding balloon, from below, a flange broke off and left it behind in the ocean as an island chain. Compare with East Asian island chains.

Fig. 11. Experiment 2002-a. Igneous mountain formation,
along the underside of a continental lithosphere.

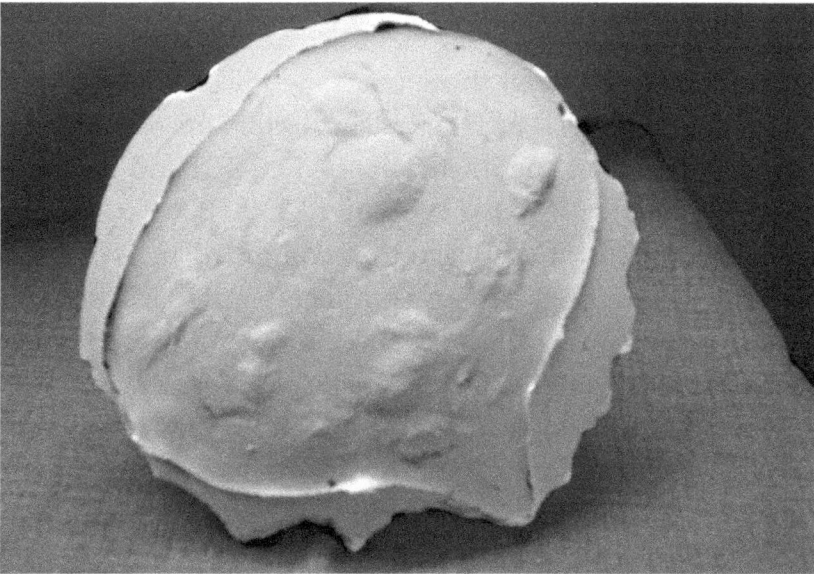

Fig. 12. Experiment 2002-b. Igneous dome formation,
along the underside of a less sticky lithosphere.

The next batch of putty was held a little drier. The mantle friction on the expanding balloon broke off the flange and abandoned it in the ocean as an island chain (Figure 10). One can compare this result with the Japanese and with other East Asian island chains.

In the year 2002 I elaborated on my 1979 balloon experiments. I wanted to know how tension, caused by balloon expansion, would tear into the underside of a slab of putty-crust. I also wanted to know how, analogically, a first generation of igneous mountains might have been injected there by rock-melt. After inflating the balloon I poured a protective plaster cast over the experimental batch of putty, for subsequent handling without having to touch. Into the mold I then cast a positive batch, representing the upper mantle, to explore where mountain ridges could have been molded (Figure 11). The expanding continental batch of putty did crack open underneath and thereby helped explain peripheral mountain formation on continents.

On the actual planet these cracks would have been filled from beneath with rock-melt and magma, from the decompressed upper mantle. After the intrusions had cooled and hardened, they would have bulged up by more rock-melt, by cooling and thickening Expansion Squeeze that continued to arrive from intermittently decompressed areas. As these materials were being cooled from above, and became sluggish, they slowed to creeping speeds and adhered to the underside of the cooler crust which lay overhead. The actual rising of mountain ranges happened somewhat "hydraulically," and it was a rather simple process. More magma was squeezed toward the flanges from under the middle of the episodically collapsing domed craton, to be cooled and stiffened. A next batch of putty was prepared drier. Adherence to the balloon skin and friction were reduced. The resulting dome-shape bubbles (Figure 12) corresponded to domes of granite which I have seen in Zimbabwe, Africa.

Episodic Expansion Squeeze.—Along with the two ancillary concepts ("Tensile Folding" and "Flanging") in the 1999 edition, I also named a third, "Relative Expansion Flow," which I now will rename "Expansion Squeeze." According to the dictum of the pre-Socratic philosopher, Heracles of Ephesus, "everything flows." Geologists have enabled themselves to think about ultra-mafic rock in the mantle being capable of "flowing," while also being somehow still "solid." And obviously magma, while it may be thought of in the asthenosphere as barely creeping, certainly flows when it escapes from volcanoes into our low pressure atmosphere. Though, as it flows we re-designate it as "lava." We also know that along spreading rifts, in the oceans, new ocean crust is being mended, amended and expanded by such melt-able and cast-able hot substances.

On an expanding planet, under a domed crust atop the asthenospheric cushion, pressures are decreasing. In the upper asthenosphere, where less pressure exists more melting of rock materials can occur. We consider the designations "astheno-sphere," and "upper mantle" as referring to the same space. These designations are frequently used differentially to apply to a variety of relationships—to the realm as such, and in relation to the mantle and to the crust.

Upon the expanding Earth, decompression happens especially in the lens-shaped upper curvature under the vaulted segment of crust—of the vaulted crust which fits less and less upon the expanding upper mantle. While no vacuums are created in the asthenosphere, it may still be thought that Earth expansion transforms domed continental crust, as well as domed seafloor crusts, into "decompression pumps."

Local decompression in the upper asthenosphere will suck rock materials toward the top, into the decompression lens-area under the domed crust. In decompression, these rock materials heat up and flow more readily as magma. Arched on the expanding magma-softened cushion of the asthenosphere, the curvature of the brittle crust is doomed to crack and sag. Some of the softer materials under the crust will be squeezed sideways.

Orogeny (mountain formation) is episodic because magma melting also is episodic and happens as long as the continental dome withstands the force of gravity. The decompression that happens to rock materials under the crust represents the first phase in the process of orogeny.

This first phase of orogeny lasts as long as a brittle crust can hold its shape, sitting upon the expanding and flattening surface of the upper mantle cushion.

The second phase of orogeny is equally episodic. It follows the melting phase and represents what generally is thought of as "mountain uplift." **This phase begins when a crustal dome collapses on its magma cushion and squeezes rock materials from the upper asthenosphere downward and sideways.**

While the earlier convex continental crust, overhead, collapses at its center, it reduces excess height in essentially three ways. **First, though this is not observable, one can assume that some of the rock-melt, weighted down by the overhead crust, is being forced back down into solid state, in the lower asthenosphere.** We probably never know for sure how much of this re-solidification actually happens.

But from under the middle of a collapsing continental crust the available magma also squeezes sideways. Some of it causes buckling along the surface of the continental craton, as also it does along the surface of collapsing seafloor plates. Some buckling in the crust may result in severe over- and under-thrusting of noncompliant strata. This means that contained under a collapsing craton, strata of brittle crust get pushed over, under, and into each other.

And finally, with extra rock-melt present, the slouching vertical weight of the collapsing crust may squeeze magma from under the center of a continent, out toward its periphery. What cannot intrude into fissures overhead will squeeze ocean-ward. It will go into the formation of ocean floors, where at some point the Spreading Rift will take on the task of mending itself.

"Expansion Squeeze" may also be visualized as corresponding to the distance that our experimental balloons, Figures 9 and 10, have crept out from under the Flange by their expansion.

Protruding bulges and mountain peaks function as climate-driven cooling elements, down toward the lower crust, to cool more quickly any rock-melt that has gotten injected there, or adheres to the bottom of crust. Thereby the foundations of mountain ranges are thickened and hardened deeper. Magma that squeezes from under the continental centers, sideways under the more easily cooled mountain ranges and flanges, adheres to and thickens the foundations. It bulges the crust along its surface and causes lateral slippage, enhanced by magmatic lubrication.

If Earth Expansion—that is, episodic continental tearing and mountain formation—were to be accepted as postulates, it can be reasoned that Earth Expansion, with its heat- and pressure driven asthenospheric dynamic, very likely also influences global warming and ice ages. It influences these perhaps more than manmade alterations and foolishness can effect. This statement is not intended to encourage more foolishness, but rather as a challenge to cultivate diligence and to seek wiser adaptations to the episodic quirkiness and growing pains of our planet —of the lovely sphere that itself keeps rolling and adjusting to a universe which appears to be expanding.

The episodic buildup of weight that collapses continental crustal domes also squeezes magmas outward under the rims of continents. Massive structures of collapsing craton crusts are wedging themselves by gravity against the mountain ranges, sideways, while more magma is cooling and thickening, and more isostatic adjustments are catching up underneath, to stabilize newer mountain heights.

Amounts of available magma are changing

In the summer of 2015, in Hamburg, I was asked to present my thoughts on Earth Expansion in my native German language. My English category "Expansion Flow," in the context of Earth Expansion, has become an unsustainable hyperbole when translated with the German verb *"fließen."* Like electrical current (G. *Strom*) flows not like a river (G. homonym *"Strom"*), but is also perceived as voltage (G. *Spannung*), so also magma exists most often just at the verge of flowing.

My earlier references even to crystalline slates in the upper mantle, as magma-flow, has turned out to be reckless speech.[29] A cacophony of languages has also overtaken the Earth sciences. I will therefore remove here the notion of "flow" from "Expansion Flow" and be satisfied with communicating something like "Episodic Expansion Squeeze."

As already stated, the terms "upper mantle" and "asthenosphere" are here being used synonymously. Beyond this, my primary aim in this treatise is structural, tectonic, and evolutionary. Future adjustments of specific local isochrones in the charts may affect our general tectonics hypothesis. It is expected that the isochrone seafloor maps will in the future undergo substantial adjustments. One can expect also that the language, pertaining to "Spread and Growth Tectonics" and to the "Eocene Transition" hypothesis, will require adjustments over time.

Simply told, there is the brittle (felsic) crust, approximately 30 to 50 kilometers thick. By contrast, the oceanic crust is only

---

29 I thank Carl Strutinski for having patiently listened to my Hamburg presentations. In the aftermath he has helped me harmonize some of my homebrewed English geologic vocabulary and my even more problematic usage of scientific German—which was partly the result of six decades of estrangement from the German language. His friendly critique has helped improve the profile of this booklet, clarifying at least some of my German-American presuppositions and intentions. Along the same vein I thank Erika Luckert for helping me make this manuscript easier to read.

between 5 to 10 kilometers thick. The asthenosphere may be understood as transition zone involving the lower lithosphere as well as the upper mantle, depending on the interaction that is under consideration.

The asthenosphere is a spherical shell, approximately 180 km thick, and it functions as a transitional zone between the ultramafic lower mantle and the brittle crust. All the transitional phases, from ultramafic lower mantle rock to magmatic flow-rock under brittle crust, and all "Episodic Expansion Squeeze" I suspect do occur in the asthenosphere. Whereas the asthenosphere is approximately 180 kilometers thick, the lower mantle adds another 2900 km of more solid depth, and it represents approximately 84% of the planet's total mass.

Earlier, for my previous designation "Relative Expansion Flow," I invoked the presence of "magma" much too frequently and in an overly general manner, so as to show the possibility of how creeping or "flowing" rock materials are being squeezed away from under the arched centers of continental cratons, down underneath the weakened peripheral flanges. My primary aim, in 1999, was to explain the presence of coastal mountain ranges in terms of igneous intrusions from below, and also their even-tual piercing through the surface of the crust. Indeed, I should have placed more emphasis on the viscous resilient condition that prevails throughout the asthenosphere and which bonds with the ultramafic mantle—which, however, in the upper asthenosphere intrudes just as easily into the cracks of the cooler felsic crust overhead.

In Plate Tectonics literature, the process of magma formation is commonly associated with the concept of "convection." Accordingly, hot mafic materials are said to be mobilized initially as rock-melt, even as deep as at the planet's core.

I can follow the logic of such visualizations, except that I personally de-emphasize processes of "convection" by which hot mantle rock supposedly rises all the way from the core up into the asthenosphere. Convection currents in the mantle may not be really needed for the asthenosphere to participate in a process of planetary expansion. General Earth Expansion is quite capable of reducing pressure in the asthenosphere, sufficiently to obtain the magma that is needed for ocean floor formation and for wherever in the system magma is needed for repair. Of course, I am not saying that convection channels in the lower mantle cannot exist. Convection processes may be active there, but I do not know anyone who really knows about these. I personally will consider their existence when convection can somehow be shown, empirically, to be effective nearer to the Planet's surface—that is, if such theoretical considerations add some solid empirical evidence.

Earth expansion enables mantle materials to flow, at least in the upper asthenosphere. And there are two conditions under which this may happen. First there is the possibility that the less dense asthenosphere not only melts mantle rock materials into magma, but that the same decompression process that melts may also loosen up some massive mantle-rock an suck it up to higher levels. Cautiously I suspect, however, that the popular notion of "convection currents" within massive mantle rock, draws far too much of its rationale from the theoretical need of explaining the spreading of excess "unwanted" ocean floors. Such postulates, about theoretical necessities seem to go too far, and against the grain of what some of us consider to be common sense.

It seems obvious that under spreading rifts, and under other types of decompressed crevices in the crust, increased amounts of magma are being produced. This also is the case under established oceanic hot spots, as under orogeny in general. **Excessive decompression and magma-melting burns deeper into the asthenosphere, loosens and siphons more materials upward to be melted.** This explains the persistence of hot rift-lines along ocean floors even after continental partner shores have gotten separated from each other by considerable stretches of floor. The process of deeper magma production, under spreading rifts, creates new ocean floor in the form of lava welds.

Then, large spreads of ocean floor may rest quietly and sturdily upon the asthenosphere. As long as no crust is strong enough to be arched or to be torn by Earth expansion, no rock materials need to be suctioned upward to be melted into magma or be squeezed elsewhere for purposes of repair.

But such tranquility is not assured forever. Earth Expansion continues to flatten the curvature of the sphere and thereby, below, it breaks open fissures into the hardened crust. Some fissures in the crust have been torn open wide, suddenly, so that ordinary magma-melt could not repair quickly enough. At Iceland, along the Atlantic spreading rift, magma-welding has built up new land above sea level. A large leak in the crust apparently has caused melting deep into the upper mantle. This hotspot appears destined to melt magma for epochs to come. But depth is not the ultimate determinant of the magma-melting process.

The asthenosphere melts most of its magma in decompressed lenses right under felsic or mafic crusts. There the measure of decompression, produced by Earth expansion, determines the amount of magma that is being "requested" or produced.

While Plate Tectonics science appeals to hypothetical processes in the mantle, by postulating convection currents for subduction, **Earth expansionists assume that the whole Earth, and not only three/fifths of its total crust, is expanding.** Expansion may indeed happen in the mantle as chemical or as nuclear reaction, or as both. For all we know, there may be convection "currents." My point is that such are really not required for subduction as long as one assumes planetary expansion. And of course, expansion could possibly happen between the core and the mantle—which, certainly, would make the planet a more lively entity than it is commonly thought of these days.

However, expansion may also happen in the asthenosphere where all building materials for the enhancement of crust are already known to be modified, transported, squeezed and are waiting there to rise for us to detect them. While I personally do not claim to know the dynamic of Earth Expansion, I am certainly willing to postulate such a dynamic and entertain a formal hypothesis, of the type that all people open to science might be willing to entertain for a while.

The asthenospheric cushion, comprised of viscous rock in transition, could well be expanding on its own, upward and all around, and thereby drive the process of Earth expansion. This asthenospheric cushion could function for our planet's growth like the vascular cambium of woody plants. It may be the layer "under the bark" where, incessantly from the greater universe, neutrinos and other subatomic particles are hailing in to penetrate. Some of them may get caught up in a creative viscous and sticky process of matter and magma-formation. This may be a hypothesis worthy of consideration. However, I personally do have neither the means to obtain samples of ultra-mafic mantle rock, nor the means to decompress such materials, nor the ability to shoot neutrinos into an experimental setup.

Inasmuch as, from various indications at the surface of Planet Earth we have come to suspect that expansion is very likely happening, we may continue to reason about the capabilities of the asthenosphere and suspect additional possibilities. Whatever else we know or do not know about Earth expansion, we know at least this: that the dynamic does not seem to reside within the victimized brittle crust where Earth expansion wreaks its visible changes.

Do the Mountains really get uplifted?

We have come to recognize that seamounts, alongside mid-ocean rifts, are cooled and hardened by ocean water over time. We are aware that mountain ranges on land are also cushioned and supported by the asthenosphere. They are created by rock-melt and by hot intrusions into crevice-molds, from beneath. The mountains rest on stratified underlay that has been cooled and stiffened from above. Mountain ranges on land are propped up by Expansion Squeeze, sent by cratons as they collapse. Nevertheless, there still seems to be a step missing in this narrative—the question about "mountain uplift" itself.

For this topic it is not necessary to abandon our familiar domain of reasoning in terms of Earth Expansion. It is not difficult to explain the actual mechanism of uplift as long as one calls the literal meaning of "lifting up" into question. Every Earth scientist knows about the presence of cratons, which characterize all continents upon this planet, and everyone knows about mountain ranges that rest on continental peripheries, or seamounts that rise alongside established spreading rifts.

Earth Expansion utilizes continental domes as decompression chambers, for melting rock materials and also for pasting magmas unto the cracked undersides of the crust. Those upside-down melting-woks in which magma is liquefied transitionally, do not "uplift" mountain ranges directly as if by muscle power. There may not be anything that literally "uplifts" mountains. In addition to magma being melted directly under diapirs by crustal rifting, rock-melt is also squeezed-in sideways. Wherever it ends up, under sea or under land, it is getting cooled from overhead.

All that a crustal dome needs to do in order to generate magma, to decompress asthenospheric rock materials underneath itself, is to withstand gravity so as not to collapse. Igneous mountain ridges pre-exist embedded in the underside of crust. They are pre-cast long in advance of their rising. In advance they have become part and parcel of the overall decompression-and-collapse cycle that Earth Expansion is managing in the upper asthenosphere. Alpine peaks, ridges, and plutons eventually pierce up into the atmosphere and function more effectively as heat exchangers than the plains. Any rock-melt that accumulates under diapirs, or magma that is being squeezed-in sideways from under plains is also cooled from above. Cooling increases the firmness of the asthenospheric foundation. Gains per annum may seem small, but they accumulate incrementally and steadily. Climatic erosion will lighten the tops of mountain ranges and carry sediments to help weigh down the domed cratons, for weightier collapsing for more Expansion Squeeze and continued orogeny in the future.

When an epoch of continental decompression approaches its end, and when the time comes for the domed crust to break and collapse once again unto its cushion of magma-melt—regardless of how thick or how thin that crust may be—some liquid materials and a variety of rock-slush will be squeezed sideways. Some will end up under mountain ranges. One should keep in mind that, over many millions of years, the continental craton has received erosion loads which have lightened the weight of the mountains and added it to the lowlands.

While a craton collapses, Expansion Squeeze intrudes into cracks and thickens the underlay. The crust has grown thicker and stronger along the bottom surfaces. With increased pressure, the crust is stiffened sideways. Each collapse of a craton-dome will prop up the flanks of mountain ranges round about. This prop-up and underlay will assure that with each orogenic episode the mountains come to rest a little higher.

Down in the lower crust, freshly opening or widening crevices are getting filled with rock-melt, are getting cooled and thereby getting underlaid with thicker up-floatable foundations.

It may be impossible to say for sure whether the illusion of mountain "uplift" resides specifically in the concept of "lifting." Everything we have written here on this subject could just as easily be conceptualized under the model of craton-collapse or adjustment to Earth expansion or curvature flattening—thus with ascribing some weaker verb than "lifting."

While we are preoccupied with watching, spectacular mountain ranges are rising. Understandably, we fail to notice how broad continental cratons are collapsing incrementally. Upon a planet, mentally maintained at steady-size, one does not suspect episodic collapses anyway. The mountains appear to rise because next to them the cratons are sinking. But of course, after each orogenic episode, cratons get loaded with sediment from higher up, as well as with an addition of hot accretions from below. Mountains are getting wedged-in to float higher.

Neither "heaving" nor "lifting" is needed to raise mountains. For accreting their foundations, at the upper asthenosphere, it is enough to cool and to thicken their bases and thereby to position them a little higher. The hardening and accretioning happens under pre-molded intrusions which, being older come to sit atop younger foundations, waiting to rise gradually, piercing through sinking crusts, keeping cool.

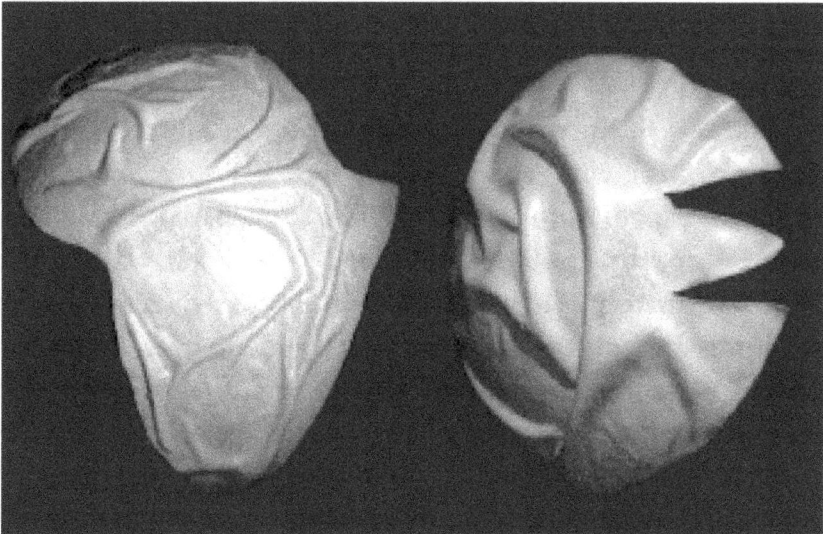

Fig. 13. Balloon skin cutouts resembling Africa and South Asia, resting on an expanding balloon, illustrating potential accruements of squeeze-materials, to be squeezed sideways as the crust settles onto the flattening curvature of the sphere. Triangular India keeps the Himalayas on mainland Asia.

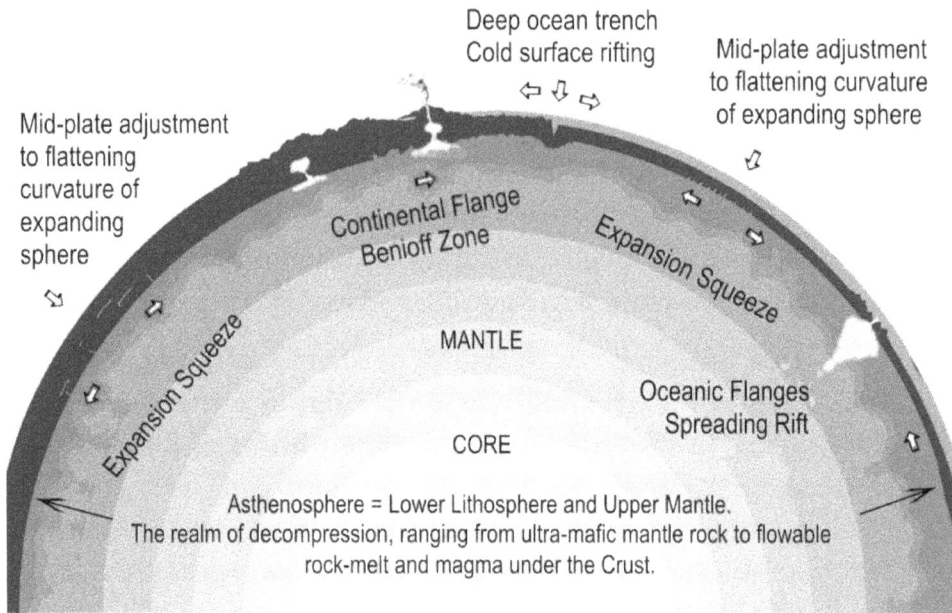

Fig. 14 : Schematic drawing, illustrating Expansion Squeeze, Flanging, Benioff Zone, Spreading Rift and Deep Ocean Trench.

Relative to the achieved height of the mountains, flatlands become flatter with every craton that collapses. Overweight mountain ranges may temporarily be squeezed, wedged in and propped up, but they are not fused tightly with crusts of the collapsing cratons. They remain standing and, from where we can observe them, they appear to rise, partly because lowlands around them must sink and adjust to the curvature of the expanding planet. To create her mountains, all that the Earth requires is a steady pace of expansion. The crust will buckle. Oceans and mountains will grow thereby, as if on their own.

Figure 13 is shown here not to explain mountain formation directly, but to illustrate the availability of rock-melt over a longer timespan of planetary expansion. It is presented to show remnant rock- and rock-melt materials from under the sinking crust, adjusting to the curvature of an expanding planet to its flattening surface.

No over- or under-thrusting along continental shores has ever been necessary for mountain uplift. **Whether shaken by earthquakes or reduced by simple erosion, thick continents, rimmed by mountain ranges had little to fear from those seven to ten times thinner ocean plates supposedly colliding with them.**

If I were a scientist, with the duty of predicting continental earthquakes, I would begin tracking the altitudes of mountain peaks on the continents, then monitor the curvature of continental cratons as well as observe changes along the flanges. Mountain ranges on land and in oceans accumulate their foundations and their ability to rise by losing heat upward, and by cooling themselves downward against the asthenosphere.

Example of the Matterhorn

Let us briefly consider the geology of the Matterhorn, a mountain in the European Alps. The peak itself represents the older portion of this mountain, and allegedly it descends from the Dent Blanche nappe of gneiss, a remainder of the Africa-Plate. Allegedly it has been pushed on top of younger ophiolite-laden Penninic nappes and on some sedimentary layers when the African continent rammed against southern Europe. But how are the older gneiss, the ophiolites and the sedimentary layers supposed to be tectonically related? Looking at the larger picture of "Spread and Growth Tectonics," along the entire stretch of the Mediterranean, one cannot find a trace of an African plate ever having gotten atop of the European plate—unless one refers to a time when Africa and Europe were together still as a single piece of crust. The eastern Mediterranean has been rifting deep since the Triassic and Jurassic. Geologic features overwhelmingly suggest tensile spreading and tearing of the crust. Continental collisions from a hypothetical pre-Pangaea era do require empirical proof before they can be admitted into the precincts of an empirical science.

During early epochs of the still underground Alpine orogeny, before Africa and Europe were rifted apart, the asthenosphere injected analogous materials from under southern European as well as from under the northern African crust. Tensile folding of the crust, caused by global expansion and stretch, tore multiple intermittent fissures into the wide syncline which extended latitudinal from the Pyrenees to the Himalayas.

Among multiple near-parallel fissures, there usually is one that has gotten torn a little wider and has been weakened a little quicker than others. Smaller fissures could be filled from the asthenosphere with rock-melt infusions. Such early intrusions were pre-formed to get raised later, as older crests of mountain ranges. The widest fissure in this broad syncline, the one that had gotten too wide to be repaired, was torn wider still, to spread the mafic ocean floor of the Mediterranean Sea. Molded mountaintops, under continental flanges, under-pasted with partly melted mixtures that contained anything from sediments to mafic ophiolites—the latter having barely escaped the fate of ocean floor formation, they all were hot-pasted to the bottom of the Dent Blanche nappe which, it is said, represents Afro-European continental sameness. After the Mediterranean Sea had rifted apart the continents, and was spread sufficiently wide, Alpine foundations were underpasted increasingly with Expansion Squeeze from under the European craton.

More than a single lifetime of comparative studies in geology will be needed to solve the Matterhorn questions which Rudolf Truempy (*American Journal of Science*, 1975) has placed on the study-desks of collision advocates. **For now, this author is confident that no continental collision was ever needed for Europe to obtain its Alps. Tensile stresses sufficed to raise them.**

5

Earthquakes, Rifts and Trenches

Earthquakes are often mentioned in arguments on behalf of an ocean floor subduction process. **Simple Spread-and-Growth Tectonics can explain earthquake dimensions as well, or probably a little better than current "Plate Tectonics" hypothesizing can.** It behooves us to share and to explain a few of these instances. Moreover, these issues can all be resolved with data drawn from sources of organized "Plate Tectonics." See Figure 14.

The 1811-12 New Madrid Earthquakes in Missouri

Over several months in 1811 and 1812, great earthquakes occurred along the New Madrid Fault, in Missouri. They raised and tilted the land, enough to cause the Mississippi River to flow backward for a time, toward the city of Saint Louis, over a distance of 150 kilometers. Based on an average differential of 23cm/km, between confluences of the Ohio and Missouri rivers, this would signify a rise of 34.5 meters in elevation. Some serious over- and under-thrusting of strata must have occurred.

The happenstance that this fault system is in the middle of the otherwise tranquil Great Plains matches the presuppositions of our general hypothesis. The crust of every large continent suffers periodic collapse at the middle of its craton, and this happens because the curvature of the crust must adjust to the flattening mantle of an expanding planet. **And because this specific seismic event happened at the middle of the North American craton, the over- and under-thrusting of strata appears to have been forced by containment in an inland region.**

Engineers in the oil business have assured me that in the American Midwest they frequently needed to drill through the same layer twice. No continental flanges or edges needed to tear open and no continental flanges needed to slip outward onto adjacent ocean floor. Crustal over- and underthrusts of this sort are a long ways from proving the subduction of ocean floor that, supposedly, would transport cold rock into the mantle to be re-melted. Nor do such quakes indicate convection currents that descend near to the core.

The 1964 Earthquake in Alaska

The earthquake in 1964 in Alaska was slightly different. It modified that region's entire southern coast. George Plafker has carefully surveyed the surface dimensions which resulted from this event.[30] He found that along the oceanfront the continental crust was uplifted several meters. Had he been a blind believer in ocean floor subduction, Plafker could have stopped his investigation right there. But fortunately he turned in the proper direction to ask additional questions. He re-surveyed some benchmarks on the continent itself and he found that the crust of Alaska had collapsed and spread outward and that the edge had slipped up a little farther over the adjacent ocean floor.

From previous sea level platforms, along the shore, Plafker could infer that such quakes were repeat occurrences. They were the expected result of what in this booklet I call "flanging." They happened on average about every 800 years.

While Plafker's observations are immensely interesting for the study of seismology, his general hypothesis is even more so for the epistemology and history of science—that is, the Fifth Dimension. Based on some type of "stable earth" paradigm he suggested that during intervals between earth-quakes, endowed with an elastic

---

[30]George Plafker. Henry C. Berg, editor. "The Geology of Alaska," in *The Geology of North America*, Vol. G-1; Geological Society of America, 1994.

capacity, the continental crust somehow springs back up to its former level. Eight centuries later it is ready to collapse again, raising the coastline another few meters.

There may be some elasticity present in the crust, due to elasticity in the upper mantle. However, any elastic back-arching of the crust, following past Alaskan earthquakes, probably never happened. **I believe that the primary mechanism for the regularity of these quakes is gradual Earth expansion—which means, the slow flattening of the mantle curvature that underlies the continental crust.** The Alaskan crust behaved very much like the crust along the Mississippi River, with one important difference. In Missouri the continental craton sagged within a constrained area, and this has caused over- and under-thrusting in the middle. The continental dome was prevented from spreading outward along its flanges, constrained by the large area and by weighty mountain ranges which have come to surround the North American continental craton.

In the preceding explanations of the process of continental flanging I have used the qualifier "to the extent that the rim holds." In the case of the periodic southern Alaskan earthquakes, the continental edge obviously has slipped and therefore flanged outward and upward. As the Pacific Ocean continues to widen along the Ring of Fire, beneath the rim of Alaska, some magma has likely been quickened by Expansion Squeeze and sea water. Continental rims have repeatedly been jarred, flanged and uplifted for tectonic adjustments. But none of this supports inferences regarding the subduction of old unwanted ocean floors by convection currents in the mantle.

Earthquakes along the Coastlines

Benioff Earthquake Zones are seismically active inclines, slanting landward and downward under continental edges. They are mentioned here, primarily, to question their significance for suspected Seafloor Subduction or Convection Currents in the mantle.

Some of the continental flanges, along East Asia, have been severed from the continent by slippage and the formation of marginal seas. In addition, seismic studies have revealed heavier earthquake activity under continental rims. The foci of the quakes occur along inclined planes that slant downward under the continental rims, some several hundred kilometers deep. These inclined "planes" of seismic activity have been named after Hugo Benioff, a seismologist who discovered them. Whereas some Benioff Zones show activity along a uniformly inclined plane, others suggest an incline along which seismic activity happens in clusters or along "stepped planes."

For the existence of Benioff Zones, a simpler explanation than convection currents and ocean floor subduction is possible. The difference between older, thicker continental crust and younger, thinner ocean floor crust implies that between the two and beneath the continental rim, there must indeed be a "slope of transition" forming.

Under the crust, looking toward land, there is not merely a downward sloping ceiling of activity but also—when the same slope is viewed ocean-ward—there also is an upward slanted ceiling of crust. Both perspectives, the down-ward and the up-ward looking, realistically visualize seismic evidence under the same continental edge.

The ocean-ward looking perspective supports our concepts of Expansion Squeeze and Flanging. Creeping slates and magmas that ooze from under the continental lithosphere are being squeezed from under collapsing continental crust which episodically adjusts to the flattening sphere. Friction-tremors of Expansion Squeeze rumble up along the Benioff slopes toward the under-side of the thinner ocean floor. Dislodged and upward rumbling crustal boulders shear, erode, flange and chamfer the steep crustal edge until, eventually, the differential depth between the continental and oceanic ceilings, which initially jutted downward into the upper mantle, is trimmed, packed, and evened out with scouring, with melted rock mixtures from Expansion Flow. Until the ocean crust can become well-established along a freshly-torn coast, a chaotic array of ophiolites may in the interval be pushed to the surface to mend the wounded coastline. Thus, earthquakes along Benioff slopes can quite easily be explained as consequences of Expansion Squeeze.[31]

The present continents are pieces of the original spherical shell, sliced by crustal tearing. If a system of cracks was present below, before the crust broke through, magma and other rockmelt would intrude to fill and mend. This process would mold future igneous mountain-peaks, while also underlay them with thickening rock-melt for gradual upward bulging—while also being cooled. Where a coastal syncline tears beyond its point of repair, it will cut through the crust. The break may widen and rift an ocean. Asthenospheric rock-flow will attempt to stabilize both flanks of a new spreading rift, regardless of whether sea or mountains are sitting overhead.

When over time the differential of an inverted coastal cliff is chamfered, along under the edge of the thicker continental crust, the process of adjustment will correspond to what presently is happening along a Benioff slope. While the bottom edge of the continental crust is being scoured by Expansion-Squeeze

---

[31]This is not to say that ocean floors are built exclusively, or even mostly, of Expansion Squeeze. However, this is how fresh spreading rifts that break apart continents are starting to spread their first ocean floors. Under every rift, even while it remains in step with a spreading ocean, fresh upper mantle materials are being decompressed and liquefied, locally. Melts are being squeezed sideways by pressures which emanate from under collapsing crusts. Under domed crusts, rock-melts become available over wide areas, by the happenstance of decompression and subsequent collapse. In any asthenospheric space that loosens, rock-melt will decompress, flow and be cooled to hardness in quick succession.

materials, the differential that is jutting down into the upper asthenosphere is being worn to a gentler traversable slope. Our analogy of "erosion and sedimentation," borrowed from the planet's land surfaces, may be inverted to visualize slow magma as getting involved in episodes such as "mafic upward falls or rapids." Younger Benioff zones produce their earth-quakes along still young and steeper edges of cliffs, along high "upward falls" or upward "creeping rapids." Speed is relative to the materials that are moving.

Benioff Zones do not need ocean floor subduction in order to rumple. To grind and bevel they quite easily can rumple in upward decompression flow, along the upper mantle.

Humankind may have difficulties imagining cliffs that facilitate upward-plunging magma-falls, or continental deep edges that invite hot inverted erosion from below—in contrast to the cold and brittle erosion by water, ice or air at the surface of the crust. But, everything that has an upside also has a low side. To approach the gargantuan mysteries of nature we must look at both.

The 2004 Sumatra Earthquake

On December 26, 2004, a 9.2 earthquake happened off the coast of Sumatra. Because it evoked the most destructive tsunami in known history, the event was frequently reported in the media. And, as always happens after geological disasters, commentators and experts tried once again a little too hard to explain what they did not know well enough about ocean floor subduction.[32]

---

[32]There are numerous useful publications on the 2004 Sumatra event. In the interest of economy I limit myself to reacting to a series of essays, published by the California Institute of Technology Tectonics Observatory, a multidisciplinary center for observing, measuring, and modeling the dynamics of the lithosphere: http://www.tectonics.caltech.edu/outreach/high-lights/sumatra/why.html. The essays are: "Why Earthquakes and Tsunamis occur in the

In its next larger context, this seismic event can easily be traced to a sort of Benioff Zone along a coastal deep trench. It certainly was not the first, nor will it be the last faulting or tsunami event in this region of the world.

On the general cookie-cutter "Ocean Floor Subduction" diagrams which audiences throughout the world are shown to naturalize geological events, upon a steady-size planet, I need not waste time or considerations. But in the context of the present book, I must respect an alleged "Benioff Zone," or "Subduction Zone Earthquake" as possibly hiding significant data. It happened right along the western edge of the Sunda Shelf, along which also runs the Sunda Trench.

In my view, this trench came into being as a tensile feature perhaps some millions of years after the Eocene Event. I surmise that the Eocene Triangle of ocean floor, exposing what formerly was the surface of the asthenosphere—which I score high among geologic data for the Eocene Transition—must have been hardening for some millions of years before it would crack away from the late Eocene shoreline of the Sunda Shelf. During the early Eocene, the Sunda Shelf was still being pulled straight south, as an integral portion of the Ninety East Ridge, still in line with Australia. At that time, none of these "plates" were in need of a trench yet.

Of course, I am looking at the 2004 Sumatra earthquake, handicapped by the same limitations as everyone else, by having incomplete data. Previous writers explained the event as a result of subduction, of the Indian-Australian plate under the Eurasian plate. Their diagrams showed how a tectonic plate was

---

Sumatra Region." "What Happened During the 2004 Sumatra Earthquake." "Using Coral to Track the History of Earthquakes." "Rethinking the Causes of Giant Earthquakes."

not sliding properly down into subduction—that it was somehow getting "locked up" and then, suddenly, was loosening and producing jolts, which were felt as gigantic earthquakes. Unfortunately, we have no altimetric measurements of what has happened over the years to the integrity of the central areas of these so-called "tectonic plates." If we had them, we could come up with better answers.

I can appreciate all references to the "speed of ocean-floor growth" between Reunion Island (near the Indic mid-ocean rift) and the Java Trench. **But I refuse to refer to the said Trench or area as the "Sumatra Subduction Zone." One cannot condone a scientific procedure that generates conclusions by merely renaming the database.** In the absence of any empirical evidence of subduction, in this nevertheless expanding universe of exploding stars and spreading galaxies, ordinary creeping expansion suffices to explain oceanic spreading—without invoking the magic of "subduction."

I am willing to admit the reality of the Trench which formerly was called the Java Trench. With political persuasion I would even consider renaming it "Sumatra Trench," though "Sunda Trench" would geographically seem fairer. It marks the western boundary of the entire Sunda Shelf.

But I do not believe that there is a significant process of "subduction," and therefore, plates moving on a hypothetical path of descent into the asthenosphere cannot "lock up." Subsequently there cannot be jerky releases. Nevertheless, an analogy from automobile technology, about brake-mechanisms "locking up" and "jerking loose" appears to be an ingenious caricature for satirizing the notion of a supposed natural process of subduction.

Nevertheless, the several long fault lines that run alongside the Sunda Trench—the one along the Andaman, Nicobar, Nias, Simeulue, Banyak, and Mentawai island chain, as well as the "secondary" fault line that breaks a sliver off western Sumatra, lengthwise, all seem extremely important. This "sliver" presents the fault that, during the 2004 earthquake caused a major portion of Sumatra to actually slide southeast. The "loosened" flange, that extends from the Sunda Trench to the said western Sumatra sliver, moved northward relative to the Sunda Shelf, which itself is reported to have moved southeastward by a greater distance. The total difference between the two sides was 20 meters. While we cannot actually see the gaps that were opened 50 kilometers down under the flange—under the island-chain-fault that held the epicenter of the quake—we can infer that somehow both faults, the one with the epicenter and the other that moved, must belong together. There has to have been some lateral looseness and sliding between them. In any case, the direction of this movement could not have been subduction. The pre-requisites for understanding this event in terms of expansion were explained in relation to Benioff zones, earlier. For context, please review Figure 13 and 14.

A consideration of deep-sea trenches is relevant for this discussion. The so-called Indian-Australian Plate, which includes the Eocene Triangle, has had time to harden and to accrete to the Sunda Plate some 42.7 Ma. But instead of becoming fused, the Trench continues to break. If we think of both so-called "plates" of the Indian/Australian and of the Sunda Shelf as adjusting their central "domes" to the flattening surface of an expanding planet, then neither of them can avoid periodic breaks. The elevations of broken edges along land-ward faults will probably be less than the the larger "plates" that broke and sank at their mid-regions. Thus, the ocean floor faults along the Sunda Trench probably needed to adjust to Earth expansion significantly less than over time the domed plates needed to sink along their middles—that is, sink unto decompression lenses which had formed in the upper asthenosphere. The edges that rose or fell along the coastal faults of Sumatra probably represented the shorter ends of the total elevation differentials that the crustal plate-domes needed to adjust.

The first flange that the Sunda Shelf abandoned, some 42.7 million years ago, was the Ninety East Ridge. The Sunda Crust, over the course of millions of years, has been breaking and adjusting, trying to abandon its western curved flange. The Andaman, Nicobar, Nias, Simeulue, Banyak, and Mentawai island chain is the flange that lies in line to let go next.

But the Trench that presently yawns between this islands chain and Sumatra is not deep enough to break off the entire crust yet—not yet. Aside from this circumstance, the bend itself provides its own reinforcement.

Along the Sunda Trench we are looking at parallel curved fault-lines which still exist as after-effects of the great Eocene bending events. Their 42.7 Ma curve, away from Ninety East Ridge, is still not stabilized. In 2004, the major Sumatra portion of the Sunda Shelf still slid in a southeasterly direction.

Tensile folding, flanging, and Expansion Squeeze are occurrences that happen under ocean floors as much as they happen under land. However, oceanic crust is thinner and therefore more vulnerable to breaking. Henry W. Menard was probably correct when he concluded early on, that the spreading ridges that he surveyed in the eastern Pacific were ephemeral.

Spreading ridges obtain their magma bulges and elevations relative to their thickness and strength—also based on the size of the arched seafloor dome which, when it collapses, exerts a certain amount of Expansion Squeeze. These factors are primarily a function of time, cooling, and of the episodic supply of flowable rock.

Areas of ocean floor crust do behave like slabs of continental crust which, as result of Earth expansion, break and settle at their mid-regions. They sag or flatten, to adjust to the changing curvature of the sphere. Under those middles, along the upper asthenosphere, continental crusts weigh, and they exert magma

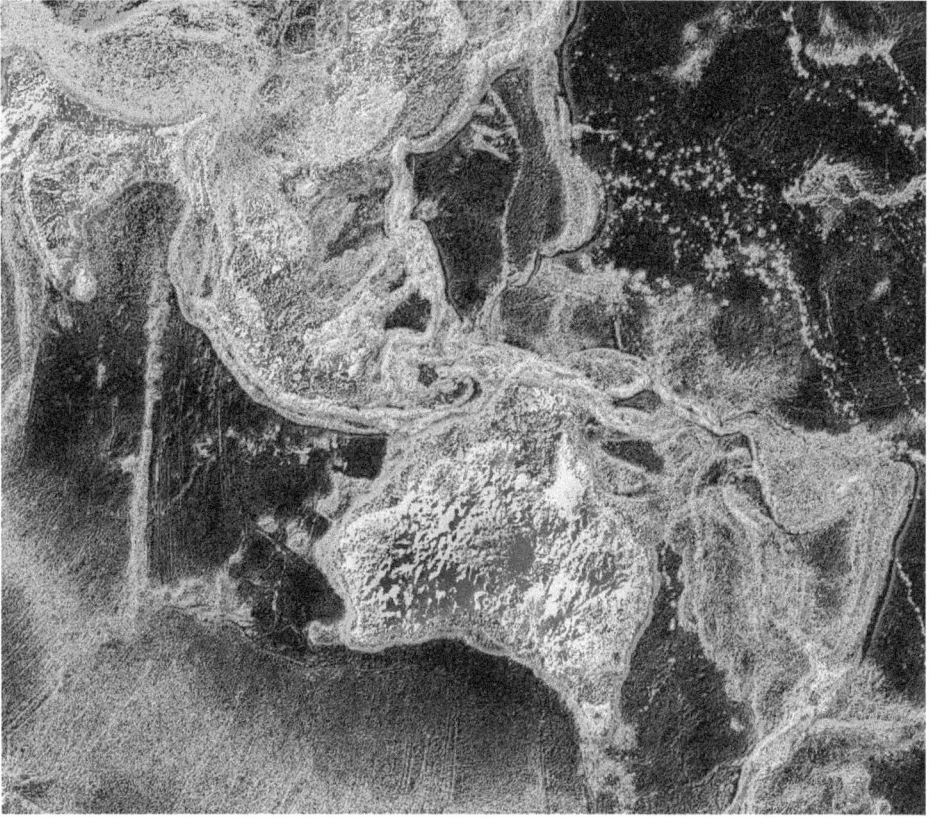

Fig. 15. Southeast-Asia with Eocene Triangle. NOAA Map: "Global Seafloor Topography from Satellite Altimetry," 1997, by Walter H.F. Smith and David T. Sandwell.

pressure sideways. In the direction of the weakest and the most agitated rims, expansion squeeze engenders increased leakage. Under any kind of agitation, even if it appears to be a pure increase of compression, random response movements to uneven compression easily contribute moments of localized decompression. Agitated and decompressed weaker spots are so mobilized for general Expansion Squeeze to intrude. While all orogenic cycles are triggered by the episodic collapse of continental crust onto the flattening mantle of the expanding sphere, prior agitation may prepare, may liquefy, lubricate, and set directions where episodic Expansion Squeeze can move.

Typical mid-ocean rifts are supplied with magma pressure that may originate hundreds or even thousands of kilometers away. The gravitational de-compression and re-compression episodes, initiated by Earth expansion, also do deliver magma from beneath to the flanks of ocean plates which are spreading sea floor along both sides of the mid-ocean rift. In addition, transform faults cut through the flanks along rifts and demonstrate how faulted segments along ridges may get spaced to achieve an equilibrium of tensions along the length of a rift.

Deep Sea Trenches

The function ascribed to deep-sea trenches, as places of ocean floor subduction, has no standing in empirical science. Such a speculation ignores all the hard-won pictorial evidence that we now have of such trenches. There simply is no evidence of slanted subduction or of scrape debris that an angular subduction process would have left.

Deep-sea trenches appear to be no more than tensile gaps. They testify to the fact that the ocean floors of our planet are crusty mafic rock. And all the ocean floors combined do, periodically, become too small to keep a growing planet enwrapped. The crusty shell-wrapping ruptures along rifts.

The deep-sea explorers, in their U-boats and submersibles, were surprised at the outset that the sediment in those suspected "subduction trenches" appeared undisturbed and of the same age as the adjacent sea. There is no scrape debris of a downward movement to be found anywhere. Instead, there are occasional deeper cracks—running lengthwise in the trenches—which serve as obvious indicators of ocean-ward tensile stress and thus, of movement in the opposite direction from the one that would be pertinent for a subduction hypothesis.

The present "Plate Tectonics" posture implies, that upon a presumed steady-size globe, the equivalent of two thirds of the Earths' crust—all ocean floors—has within less than 250 million years been recycled through such undisturbed "subduction trenches," without leaving a trace. —Truly! In this world of story-tellers, who is trying to teach a faith in miracles here?

It is of course understandable, that anyone who truly believes in a process of ocean floor subduction would begin searching among cavities and ravines near enigmatic coastlines—as well as among seismically active, so-called "Benioff Earthquake Zones." But as I have already explained, Expansion Squeeze from half continents away, weighed down by crustal adjustments to Earth expansion, can account more easily for earthquake inclines in the lithosphere under continental rims, than anything else that anyone has mentioned so far.

Meanwhile, the continental crusts continue to flatten. The new magmatic ocean floors which grow onto the flanks of continents will harden, cooled by ocean water. The ocean floors themselves are spreading. They will grow to the size of larger plates that in time must also sink their middle areas to adjust to the flattening surface of the expanding planet. As they settle in the middle and crack from below, magma intrudes and mends the wounds. Like a felsic continental crust, so also the mafic ocean floor will episodically collapse and exert magma-squeeze pressures from under its mid-plate dome toward its flanges.

But what will the landward portion of a half-ocean plate do to cushion the effects of adjustment-drops that happen at its domed middle? It will attempt to pass some of its stresses outward and, depending on the ocean-plate's bond with the continental shoreline, the landward stretch of ocean floor may crack open a fresh trench, or it may deepen an existing one. As a dome collapses along its middle, the crevice or trench will likely crack open along shorelines, from the surface of the ocean floor down toward the asthenosphere (Figure 14).

Deep in the ocean trench, along the underside of the water-cooled crust, cooler temperatures will promptly invite a patch of magma reinforcement to be pasted on underneath, cooled and hardened. While the ocean plate is collapsing far out at sea, trenches near the continent will crack deeper.

By contrast, mid-ocean spreading rifts are crevices that crack through from the hot asthenosphere below toward the cool ocean water overhead. Magma is welded along those seams as lava, to seal the widening rift at both sides. Expansion squeeze arrives from under the adjoining collapsing domes of ocean plates. Extra rock-melt accumulates under the rims. It mimics the process of mountain formation along continental synclines. Rock-melt plugs and seals incomplete cracks from below. It pre-forms seamount peaks by magma injection and gradual episodic rising. As it happens with continental plates, so the oceanic floor plates will "schedule" their episodic uplifts, also their periods of stagnation, in accordance with cycles of crustal collapse and crust re-stabilization.

Much of the magma that goes into ocean floor construction, along the mid-ocean spreading rifts, is probably melted locally, right under the rift. By contrast, much of what goes into the rise of seamounts, upon rims that flank the rift, may have seeped in sideways from under collapsing domes of ocean floor crust. The larger Earth expansion process, on land and in oceans, adjusts the asthenospheric pressures—vertically by collapse of crustal domes and laterally by Expansion Squeeze.

Trenches and mountains both do function as cooling elements. They thicken and apply rock-melt patches along the ceiling of the asthenosphere. In crevice-molds they harden the intruded rock-melt as soon as the process of cooling catches up with the intrusions. Under trenches in the asthenosphere, a buildup of magma creates reinforcement. It builds an inverted dam that patches and stabilizes the crack along the bottom of the crust, while seawater still cools and thickens nearby the foundations of previously filled molds of potential mountain crests. The next expansion break, in a trench, will need to crack deeper.

Upon a growing Earth, there are no lasting mechanical solutions. If we were living on a planet, built by human engineers, whose minds are set on raising mountains by bulldozing them into place and shape, sideways as if by subduction, I reckon that our coastlands would look much more chaotic than presently they appear.

Aside from providing serious mythological support for heavy earth-moving equipment industries, the Convection Current and Ocean-Floor-Subduction hypotheses do not make our understanding of episodic mountain formation any easier.

It actually has become difficult to find pure empirical time-specific data pertaining squarely to physical orogenic sequences. Most scientific reports on that subject matter, today, are already predisposed to continental collision-and-subduction episodes which, according to this author's personal perspective are highly speculative. The task to rummage through these contaminated data will require another generation of Earth scientists to sort out.

Part Two:

Three Oceans are One

6

The Atlantic Ocean

The nearly parallel coasts of the Atlantic Ocean tease scientific minds to attempt a realignment of its coastlines. The Dutch geographer Abraham Ortelius, noted the matching shorelines of the Atlantic Ocean way back in 1596—and so, in all likelihood, has every cartographer since. On the basis of fairly regular shorelines, it is easy to see how the arc of northwest Africa must have come from the bight of North America, and how the knee of South America must have been in the Gulf of Guinea. Two regular Jurassic slivers of ocean floor are situated exactly where they would be if the Atlantic underwent symmetric rifting and spreading. The Jurassic strips run along both sides of the North Atlantic curve, indicating deep and ongoing ocean spreading there by 150 Mya.

The Atlantic Ocean has a clearly defined mid-ocean rift, along which east-west expansion happens. Transform faults resulting from north-south tension are spaced regularly along the ocean's length. During the Upper Cretaceous, the Atlantic rift tore into Baffin Bay, west of Greenland. However, its main northern spread, east of Greenland, was opened later, during the Paleocene. Iceland was extruded from that main North Atlantic rift, at a point of extra stress between Greenland and Europe. There, the Atlantic mid-ocean rift has erupted creating land above sea level. A triangular peninsula, Greenland, now splits the "fork" of the North Atlantic. If one ponders continental separations in association with deep-rifted oceans, then "continental shelves" should be counted with continental crust. By that reckoning, Greenland does not appear as a large island, but as a triangular peninsula, somewhat like India, Arabia, and Sinai, with its shape torn from the south.

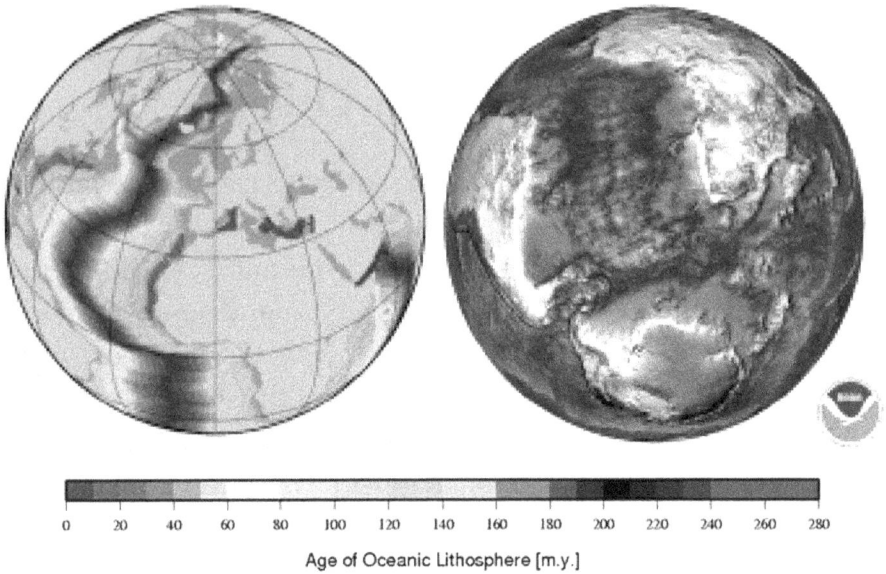

Million years – this scale applies to all NOAA Isochrone hemispheres

Fig. 16. NOAA Isochrone-hemisphere for North Atlantic;
Topo-hemisphere for South Atlantic.

Two strips of ocean floor from Jurassic times run along the east and west coasts of the North Atlantic (Figure 16 left, and Figure 17). Somewhat later, during the Lower Cretaceous, the South Atlantic began to tear open as well. It loosened South Africa from the toe of South America (Figure 16, right). Evidence of the Eocene Tectonic Event, in the Atlantic, is limited to two main areas—Middle and South America. The Panamanian Isthmus and the West Indies Ridge both were pushed northeast-ward, and when this occurred their impact pressures continued unto the Atlantic Ocean floor.

Their disfiguration suggests that, during the Eocene, some differential movements occurred between the South- and North-American crusts. A patch of Paleocene ocean floor was pinched off into the Atlantic, above the Vema Fracture Zone, at latitudes

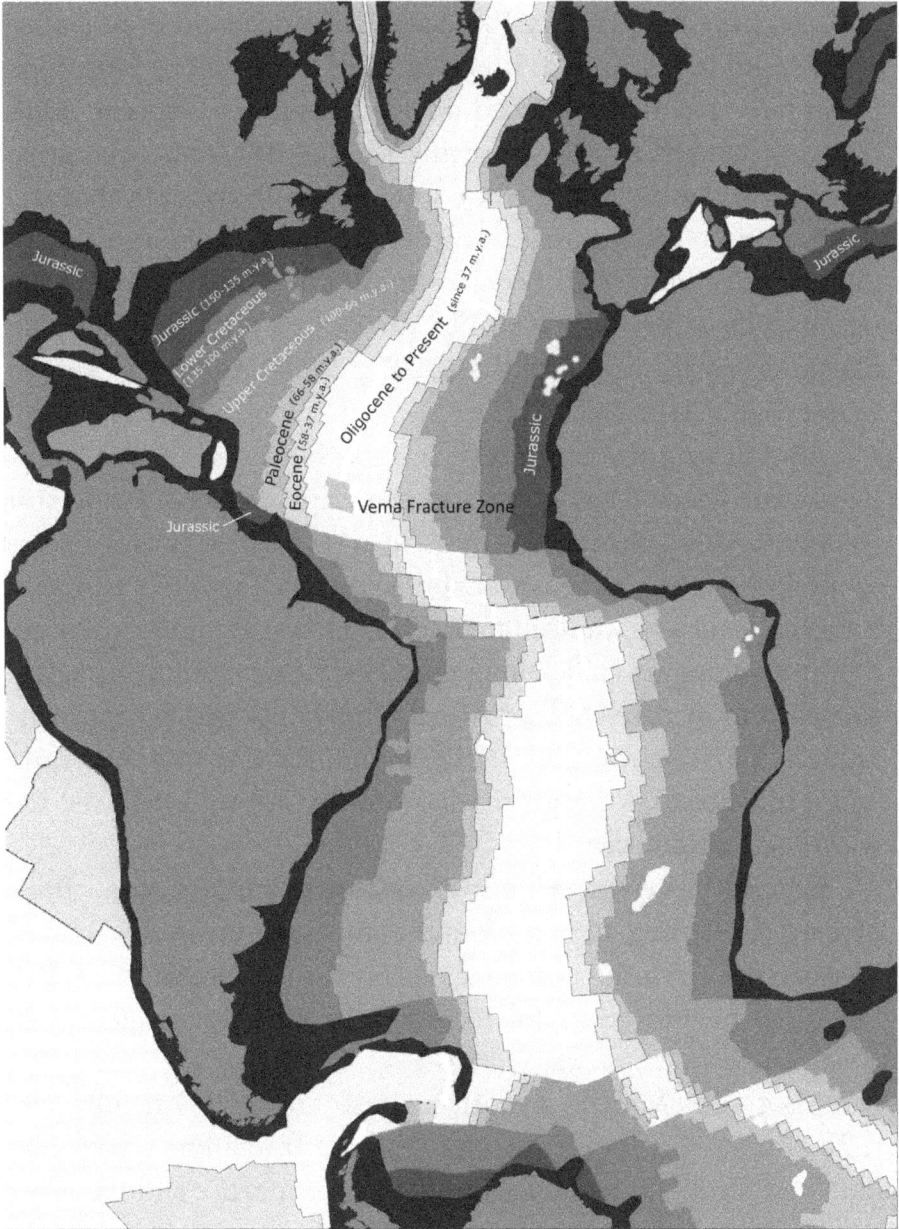

Fig. 17. The Atlantic Ocean. Paleocene patch in the "Vema Fracture
Zone." Drawn after UNESCO Geological World Atlas, 1988.
Reprinted from the author's 1999 edition.

that also mark the West Indies bulge. Even though the map-makers show the primary Paleocene strip near the West Indies bulge regularly rounded, it is clear that the patch of Paleocene ocean floor has been pinched off from regular Paleocene areas by a sudden jolt of Eocene rifting (Figure 17). This process of separation continued into the Oligocene. It appears that the southern tip of the Jurassic curve, at the North American side of the Atlantic, which had been severed earlier and pulled south, was by that same Eocene South-American jolt twisted and displaced eastward by more than 2200 kilometers. It now can be found east of Trinidad.

Also during the Eocene, at the southern end of the Atlantic Ocean where Africa's tip and the Cape of Good Hope still were embedded during the Jurassic, a portion of Lower Cretaceous floor was pushed eastward by Antarctica. During its partition from the Americas, the Antarctic Plate turned counter-clockwise, tightly shearing its "tail" along the Cape Horn of South America, eastward in the direction of Africa. This happened during Antarctica's twist southward, while it took the place of Australia and while South America made room by jolting northeast toward the Florida area.

Beyond basic ocean floor chronology, JOIDES ocean floor drilling (ODP, Leg 165) has found some surprising amounts of volcanic ash from the Eocene all across the Caribbean Sea. ODP, Leg 171B has, in addition, recorded an Eocene change of sea level.

All across Middle America, these anomalies fit into the time frame of a larger Eocene tectonic transition which has unfolded primarily in the Pacific, the new Antarctic and the Indic oceans. Here, along the Atlantic, the effects of that event are related to the sudden northeasterly shove of South America. That same movement caused the Panama isthmus and the West Indies Ridge to bend northeastward. Leveraged by the jolt that it received from South America, the agitated Florida area and the southern portion of North America were pushed some distance westward. Southwestern North America thereby had a sliver of California pushed over the Eocene coastal rift which now, underneath land, is known as the "San Andreas Fault." The sliver of land was pushed into Pacific space which a little earlier had been vacated by Antarctica.

Antarctica, during its counter-clockwise swing, pushed the Falkland Plateau, the Scotia Ridge, and the Sandwich and Orkney islands eastward by the velocity of its movement and the power of its tail. It used its "tail," which formerly cushioned the Great Bight of Australia, like an iron heel or wedge and destroyed some of it in the process. The eastward push of Cretaceous ocean floor, toward the southwestern coast of Africa, apparently reduced tension in the western Mediterranean, up north. More space for new ocean floor to form was created there after the Eocene.

Prior to the Eocene Transition, the same global expansion pressure that opened the Mediterranean gap also loosened the Arctic area and lessened, for a while, the longitudinal pull that lowered Southeast Asia. During the Cretaceous, also the Central Plains of North America were lowered by "tensile folding" for inundation, to form a large shallow sea.

This process has opened the Jurassic Gulf of Mexico and thinned out the entire Middle American region. The area, as part of the circum-global belt of continents consisting of Australia, Asia, North America and South America, has been stretched over both poles. The same tension has also lengthened South America southward. Today, South America is 400 to 500 kilometers longer than it needed to be at the beginning of the Cretaceous to accommodate the size of Africa.

The entire planet expanded faster in the south than in the north. It therefore tended to tear triangular continental shapes which pointed south. The three continents of Antarctica, South

America, and Africa all tore themselves in pointed curves away from the Bight of Australia. Other southward pointing continental triangles on the planet are India, Arabia, Sinai, and Greenland.

Then, in preparation for the great Eocene tectonic event, already during the Paleocene, South America, Antarctica and Australia began cutting their joint "liberation curve" a little ways into the Great Bight of Australia. During the Eocene, these three continents effectively separated. Antarctica then pushed its pre-carved tail between the two other continents and continued to tear itself away from South America.

The Mediterranean Sea and the Black Sea must also be mentioned in connection with the general worldwide stretching and slicing as it happened in the Atlantic. The eastern Mediterranean and the Black Sea were both being pulled open southward during the Triassic and Jurassic, while sitting above the toe of South America, Africa still had a foothold on the overall circum-Pacific belt of continents. When that toe-hold was lost during the Lower Cretaceous, expansion in the northern stretches of the Eurasian and African Mediterranean ceased. The western Mediterranean Sea was pulled open later, northward, while Europe and Greenland were still pulling at each other, trying to break the Atlantic Ocean open toward and into the Arctic Ocean.

Early on, when the round Pacific began to open with a triangular patch of Jurassic ocean floor, and when Antarctica still was being carved from the planet's virgin crust, the global pattern of tensions for the remaining crust was readjusting and changing. Some 150 million years ago, the North Atlantic cut southward along a westward curve. Approaching the equator, it turned southeast and then south, at a near right angle around the knee of South America. Thence it sliced a more-or-less straight line south and westerly.

All the while the Pacific gave birth to its legitimate round continent, with a "tail"—or shall we take it as an "umbilical cord" in honor of the emerging placental mammals at the time? The Atlantic rift gave us a similar amount of curvature. By way of changing its direction east-ward, it tore South America away from Africa; by way of cutting the shape of a rather elongated and irregular "*S*," the Atlantic has become the most linear among our oceans.

The elongated shape of an *S* of the Atlantic Ocean in its entirety, suggests that natural laws involving global expansion, gravity, cohesion, tension, and curvaceous tearing were interacting in the formation of this and other oceans. These same laws have been interacting for some time to carve the shape of a "9" for the Pacific. This happenstance further suggests that the somewhat brittle crust, covering an expanding sphere, necessarily tends to tear along curves. Some curves are compromised by changing into reverse curves, angles, or approximate straight lines. It is Earth expansion pressure, from within, that affects the overall tensile pattern among the vaulted continental crusts. It is the expanding planetary sphere, however, that generally has shown a preference for tearing into the underside of its lithosphere mostly in curves. The modification of overall tensile patterns in the crust, by the very fact that tearing was under-way, has itself changed curves into angles and straight lines into curves.

7

The Indic Ocean

Perhaps because of the somewhat irregular array of the Jurassic ocean floor data, in the Indic, are we left to assume that this body of water is the youngest among the three oceans. After a relatively small gap had been torn for the eastern Mediterranean Sea, during the Triassic, the first major oceanic rift that opened became the Pacific. The Atlantic opened next and the Indic perhaps last. Fortunately, the exact Jurassic ages, of the three oceans, are not essential for this essay. We began in the previous chapter discussing the Atlantic Ocean, because it is simplest to understand and to explain. We turn to the Indic Ocean next, because it contains the second most important data regarding the Eocene event. Thereafter, we will engage ourselves with the Pacific, because its evolution is central for understanding the Eocene Transition in its entirety, for the placement of Antarctica and the opening of the Southern Ocean.

When the round Pacific had been spreading for some twenty million years in a westerly direction, probably to ease a north-and-south tension, a southward curve was torn for the North Atlantic, apparently to ease a general east-west tension. At the knee of South America, the Atlantic gap has added a near-right-angle to complete something resembling the letter *"S."* The veering Atlantic rift tried to adjust to what was happening differently with regard to crustal tensions halfway around the planet. And finally, reacting to both the Pacific and the Atlantic rifts with a genuine linear rift, the Jurassic Indic broke open longitudinally, northward, along the eastern flank of the African continent. It was tearing parallel to what now is known as the

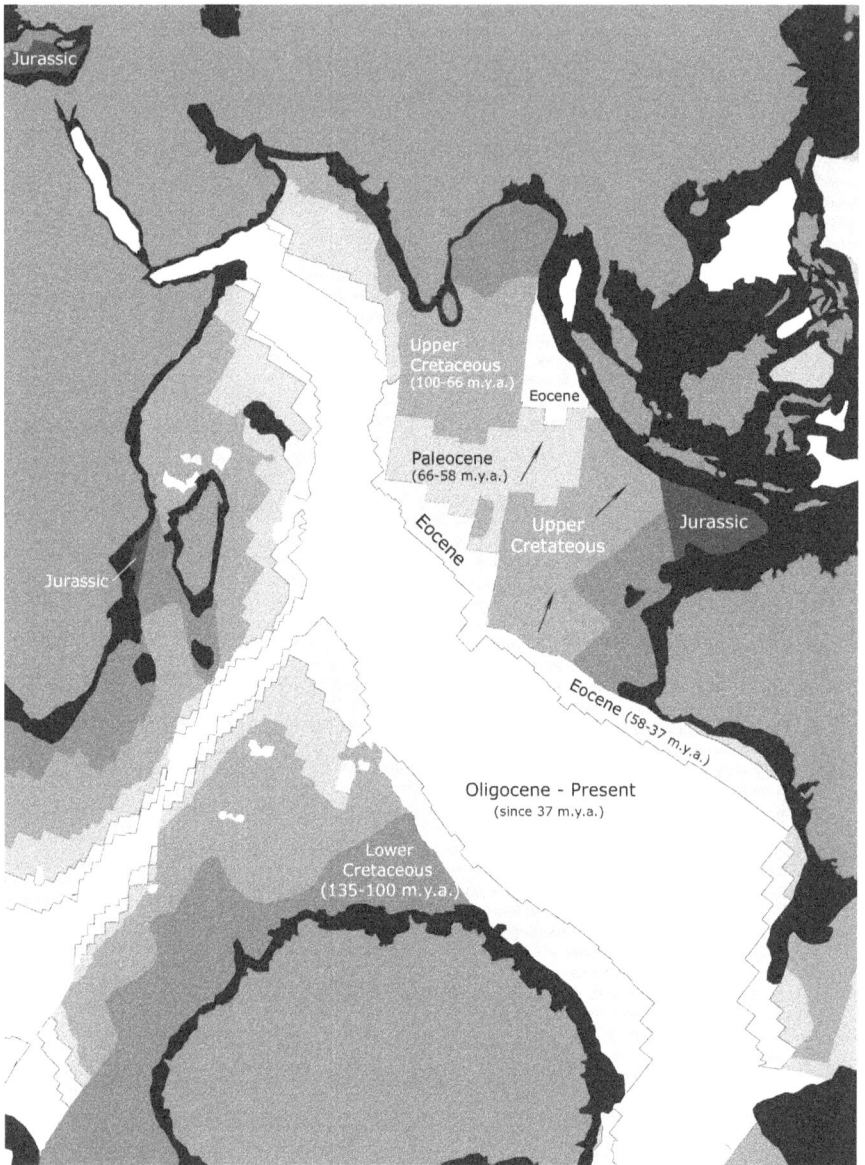

Fig. 18. The Indic Ocean. Drawn after the *UNESCO Geological World Atlas*, 1988.

African Rift, which presently still tries to flange away from East Africa. The first Jurassic Indic rift also ran parallel to what has become the longitudinal Ninety East Rift. But then, during the Paleocene, in the Indic Ocean, a significant transverse spreading rift ran east-west for a while. Relative to the faster expanding

southern hemisphere, the southern edge of South Asia probably retreated a little northward during those epochs of ocean
formation. **The present dominant spreading rift of the Indic
Ocean, which runs from the southeast to the northwest, has been
cutting diagonally only since the Eocene.**

The older straight Jurassic rift, parallel to the eastern coast
of Africa and the Ninety East Rift, was left in doubt for a while.
The makers of the UNESCO Indic Ocean Isochrone-map forgot
to indicate the Jurassic portion in the Mozambique Channel.
However, the missing information could easily be retrieved
near the edge of their Atlantic map. Meanwhile, this gap in
information has been filled in, on the 1996 NOAA-map titled
"Age of the Ocean Floor." The blue Jurassic rift that ran along
East Africa is indicated there clearly now—almost as if with a
vengeance.

During the Lower Cretaceous the island of Madagascar
was still close to Africa. Then, during the Upper Cretaceous,
this large island and its plateau slid southward. For some years,
this perception has been disputed by some Earth scientists. But
there no longer remains a need for doubt. The scar along which
Madagascar slid southward is now clearly visible on the new
NOAA Map: Global Seafloor..., which shows the topography
of the ocean floor in remarkable detail.[33] Chronological reconstruction of the eastern Indic Ocean still is a subject matter of
considerable controversy. My own interpretation differs from
solutions which were offered by other proponents of Earth Expansion. During the Lower Cretaceous period, the Ninety East
Ridge was still combined with the Chagos Laccadive Plateau.
Together, these scars on the ocean floor reflect the direction of

---

[33]NOAA Map: "Global Seafloor Topography from Satellite Altimetry,"
1997, by Walter H.F. Smith and David T. Sandwell.

the original Jurassic rift that severed Africa from Southeast-Asia. This older rift has forked in the north to slice the triangular shape of India. One branch opened up what today is known as the Bay of Bengal, and the other branch veered northwest to separate India from Arabia and the Horn of Africa. The northwestern branch has become the primary rift, especially in the northern portion of the Indic Ocean. It is presently engaged in opening up the Red Sea.

Even though we already have discussed the process of mountain formation in general, some mountains of Asia deserve special mentioning at this point—even if for no other reason than for the fact that Alfred Wegener, and some of his followers in Plate Tectonics, have in all likelihood already been saying a little more about mountain ranges and the Indic Ocean than they knew.

The subcontinent of India never could have drifted across the ocean that now bears its name. This much becomes clear upon contemplating the topography and the chronology of the rifts and ridges which now define the Indic Ocean. The fact that an independently mobile Indian continental mass is still being considered by avant-garde Plate Tectonics theorists, may forebode a whirlwind of controversy for this booklet. But I happen to be convinced that **India was never farther away from the continent of Asia than it is today—though, at one time it surely was much closer to Africa, Arabia, and to Southeast Asia than it is now. Moreover, I also believe that India's crust only "appears" to be diving down under the Himalayas because that crust has suffered some structural collapse under the weight of the rising and spreading diapir.**

The global tensile forces that tore the outline of India are the same that gave similar triangular shapes to Greenland in the North Atlantic, to Arabia, and to the Sinai Peninsula at the northern end of the Red Sea. One should consider land masses in terms of their general cohesion—and not by the arbitrary criterion of sea levels where a few meters difference in altitude can classify continental crust as sea. Greenland is a peninsula of approximately the shape and size of India. While a small cut near Spitzbergen indicates that the Atlantic rift has been making inroads on the Arctic Ocean, continental shelf and crust does still surround this Arctic Ocean everywhere else.

In any case, all these triangular peninsulas were torn from the original Earth crust by the same force of above-average Earth expansion in the southern hemisphere. But there is, of course, some significant difference between India and Greenland in their relation to neighboring continental crust. Along Greenland's northern edge, there is no massive continent that could flange for itself a mountain range comparable to the Himalayas. Nevertheless, it seems noteworthy that the two most linear-tending oceans, the Atlantic and Indic, aligned initially south- and northward and have both rifted a V-shaped crust at their northern ends.

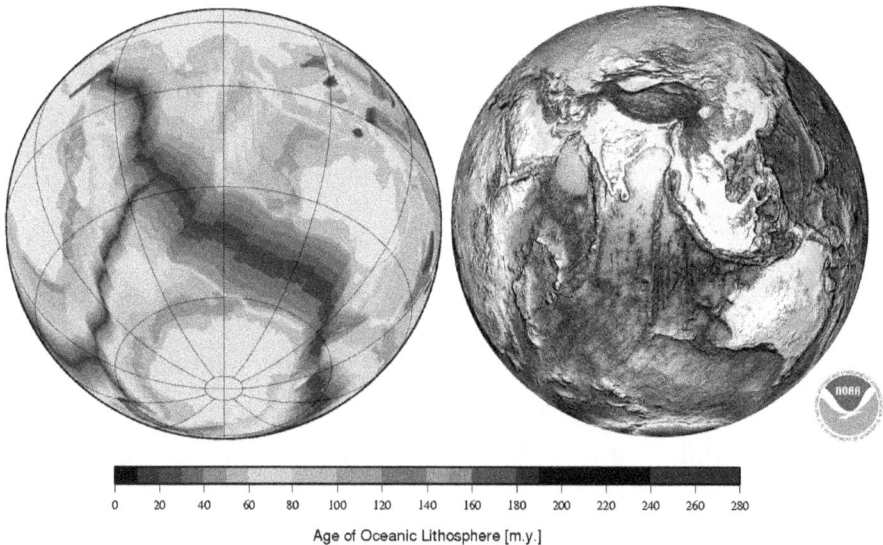

Age of Oceanic Lithosphere [m.y.]

Figure 19. The Indic Ocean. NOAA Isochrone- and Topo-hemispheres.

Why did one or the other of these rifts not tear its V upside down, as a Δ down south? The answer is clear from observations already mentioned. What the Atlantic Rift did to Greenland, what the early Indic Rift did to India, and what the Red Sea presently is doing to the Sinai Peninsula, could not have occurred down south in the same manner. "Down under" was no broad continent left into which triangles or pointed slivers could be sliced. The preponderance of Earth expansion happened in the southern hemisphere. Moreover, the two continental units—Antarctica and South America—that remained of the global belt, were slivers that themselves pointed south—albeit, pointed in the same direction as Greenland and India. But they were slicing their wedges as final curves into the larger curve of what became the Australian Great Bight.

While the three great oceans—the 9-shaped Pacific, the *S*-shaped Atlantic with a V up north, and the I-shaped Indic with a V up north—were spread open, the narrow remainder of the longitudinal belt that encircled the globe withstood the general expansion stress until the Eocene. The continents that made up the last circum-global belt were the same ones that today line the famous Ring of Fire. They included the Americas and Asia, all the way to Australia—the latter was the continent which until the Eocene functioned as the very "buckle" of this last longitudinal global belt. In the last phase of stretching, three continents still held on to each other along the Great Bight of Australia, clear into the Eocene.

Many a creative vision appears to be rooted in childhood. And to that effect, four geographical features have impressed themselves upon the mind of this writer, back in a one-room village school. These were the "Dog of Scandinavia," the "Boot of Italy," the swerving "Tail of Antarctica" that suggested motion at first sight, and the severely twisted "shape of Southeast- and Austral-Asia." While the first two served playful imagination regarding Platonic prototypes in geography, the other two awakened a primitive curiosity about a planet that appeared to be alive and in motion. I shall now attempt to satisfy this youthful curiosity concerning the fourth of these geographical wonders.

In this treatise, the broad sequence of Eocene geological events is referred to repeatedly in relation to the formation of the Pacific and Atlantic oceans. This is necessary in order to place the features of other oceans in the context of the larger taffy-pull which happened down along the asthenosphere. Now we have arrived at a point where the shape of the entire Southeast Asian continental configuration is an issue. How did Southeast Asia get to be so twisted?

The Paleocene epoch saw the beginning of small rifting at the Great Bight of Australia. This advance tearing during the Paleocene prepared the way for the larger breakaway that soon would decisively alter the face of the Earth—namely, the severance of Australia from Antarctica and South America during the later part of the Eocene. Serious rifting also had begun between Australia and New Zealand during the Paleocene.

The overall tensile pattern of Southeast-Asia can be seen, still, all the way from the stressed continental patches south of Tasmania and New Zealand, and also northward into China, to mountain ranges which were bent to run southward, as far north as the Qilian Mountains. The same tension that stretched Southeast Asia southward in early Eocene times also pulled mountain ranges along the eastern edge of the Tibetan Plateau into a southward alignment with the mountain ranges of Southeast Asia.

Two sets of ridges are visible on the present sea floor of the Eocene Triangle, west of Sumatra (Figures 15, 18). A set of narrow ridges runs north and south, essentially parallel to the Ninety

East Ridge, and another bulgier set of ridges. The narrow lines look as though they began as fissures of tensile folding along the underside of the Sunda crust, atop the asthenospheric global belt. A somewhat broader set resembles the basement contour of the island of Sumatra which, during the Eocene movement, was twisted northeast-ward. Both of these are the remains of rock-melt ridges, initially intruded from below into crevices and cavities. Their different shapes show them as intrusions with differt overhead and differential cooling over time (Figure 15).

The narrow ridges that stripe the "Eocene Triangle" longitudinally were formed by Tensile Folding. The sharp, steep ridges of these intrusions were probably broken off by the same process of tensile folding. They were probably left stuck in the underside of the continental Sunda crust. While in early Eocene times the Ninety East Ridge represented the western edge of the global belt, the area of the Eocene Triangle represents a bared surface of the asthenospheric stratum of this belt that formerly stretched and drowned the Sunda Shelf. Now its surface has hardened. The crust that originally rested on this asthenosphere was loosened by the steady increase of tensile folding which the entire global belt experienced over time. The increase of tension produced a constant crawl and shearing on the part of the crust, while both land and water cooled the mafic rock from above.

A deep scrape across the asthenosphere along the Australian Great Bight, to be explained in the next chapter, severed the global belt. The loosened crust of East Asia, prodded by Australia which itself was on rebound from the cape of South America, could thereby leave some of its former asthenospheric bases, and then slide sideways across the foundation contours as well as across a set of narrow longitudinal stretch-ridges. The intruded ridges were broken off at half height and are probably still stuck in the underside of the Sunda crust. The moving crust so slid across its asthenospheric basement which, at the time, might still have been marginally elastic.

During the Eocene Event (ca. 42.7 Mya), the bottom of the sliding Sunda Shelf apparently was still elastic enough for its western contour to be bent northeastward without much splintering. On the other hand, to this day, even though the Sunda Shelf is now curved, the brittle-cooled transverse fault under Sumatra still breaks and flanges lengthwise.

Thus, along an Eocene seafloor it is still possible to discern the former footprint of Sumatra, from a time before the bending occurred. Differential cooling of an old asthenospheric melt-zone, under land and water, has enabled the island of Sumatra to leave its older contoured footprint at the surface of its previous asthenospheric basement.[34] These circumstances are rendered visible on the newer topographical maps of the ocean floor. Enhanced by satellite altimetry, contours and elevations now reveal some excellent detail (see Figure 15, above).[35]

While the logical framework for this geological interpretation is for the most part based on magnetic isochrone data, made available by new maps of the ocean floor, the altimetric Satellite maps enable us to visualize the seafloor now better than before. **The Ninety East Ridge and Broken Ridge together, the bent ridge along the Sunda crust, plus the bare asthenospheric ocean floor of the entire Eocene Triangle, are significant paleoscars that testify to the Eocene Transition for the Indic and southwestern Pacific region.**

---

[34] Because we do not know the precise place and surroundings where Jonathan Dehn's date was obtained (42.7 Mya), we expect that in the future, when we are ready for finer contextual dating, there will be some differences among asthenospheric periods of cooling to consider. See also Footnote 37.

[35] Similar footprints of an island crust, partially outlined at the asthenospheric seafloor north of Madagascar, can be found on the same altimetric satellite World Map. This author became sensitized to such footprints as a consequence of his balloon experiments. Compare Figures 10-11, above.

Verification of the great Eocene Transition can also be found along the eastern edge of the Southeast-Asian stretch. The Philippine Sea opened its first spreading rift during the Paleocene. It ran north and south and, in general, it paralleled the inclination of Southeast-Asia in its earlier stretched direction — this means, in line with the eastern edge of Southeast-Asia, as well as parallel to the Ninety East Ridge along the western edge.

During the Paleocene, the bond between Australia and the tip of South America was still intact. But at the moment when the Eocene tectonic event loosened Southeast Asia and allowed it to pull north and to veer eastward, the direction of this spreading rift, in the Philippine Sea, changed to match the diagonal direction that the spreading rift of the Indian Ocean effected then over the larger realm. The Philippine Sea opened, and the Mariana Ridge with its islands was being squeezed twice from the south. This means, pushed from the south while the entire East Asian mainland abandoned its large flanges — Japan, Kamchatka, and the Aleutians. Relatively speaking, it can also be said that the Asian mainland experienced some westward circumferential slippage. Along the coast of eastern Asia most of this happened during the Eocene. It was the easiest way, there, for felsic crust to adjust to Earth expansion.

In the Indic Ocean, the Paleocene and Upper Cretaceous portions, that slid northeastward alongside Australia, are bounded in the south by the Broken Ridge, which appears to have been broken and pulled along at Australia's west side. The Broken Ridge, too, represents a fragment of the circum-global belt. It appears to have been a southward extension of the Ninety East Ridge. New Guinea and Australia together were part of the continental unit that stretched and thinned the entire assembly of Indonesian islands, southward. Elongating and stretching, the global belt produced the tension that introduced straight lines in the Indic Ocean, some of which still run longitudinally and have left the bases of linear intrusions on the hardened surface of the asthenosphere (Figure 15). This was the same circum-global tension that also tore open the Gulf of Mexico during the Jurassic, and which gradually elongated both Americas longitudinally along the eastern edge of the Pacific.

The Eocene Tectonic Event in the Indic appears to be dateable. "Leg 121" of the Ocean Drilling Program (ODP), which reports on a research voyage to the Indian Ocean by the drilling ship JOIDES Resolution, listed its objective as drilling into the Ninety East and Broken ridges. It further set out to investigate "the evolution of the Kerguelen/Ninety East hot spot and problems related to the rapid northward movement of the Indian Plate during the late Cretaceous and Tertiary."[36] The scientific reckoning, after drilling into Broken Ridge, was as follows: the "results indicate >1,000m uplift in response to a Middle Eocene rifting event. The short duration of the rifting event (3 to 7.5 million years) and low present-day heat flow suggests a mechanical rather than a thermal mechanism for uplift."[37] The specified brief span of time, of 3 to 7.5 million years during the middle of the Eocene, attracts our attention. It matches our Eocene Transition hypothesis quite nicely.

Jonathan Dehn, a tephrochronologist who participated on that voyage, echoes the same conclusion to the effect that in the Mid-Eocene "the Ninety East Ridge and the Broken Ridge were separated from the Kerguelen Plateau and Hotspot, for the formation of the South East Indian Rift."

---

[36] There is a point of encouraging news contained in this recorded research goal. Those scientists who proposed this expedition, at least saw problems with the Wegener theory of India's association with the fictitious Gondwanaland and its subsequent collision with South Asia. If sea-floor drilling is required to dispel this impossible notion, then by all means, let us drill!

[37] http://www.odp.tamu.edu/sciops/Leg_Summaries/Leg121-140/ Leg121-html.

We may tentatively put brackets and question marks with most explanations that link the aforementioned ridges to the Kerguelen Plateau—it is enough if, for the time being, these notations are kept bundled faithfully under "ODP Leg 121." Nevertheless, the Eocene Tectonic Event, in the northeastern Indian Ocean, is suggested by the eventuality that the Antarctic plate, with its triangular, older ocean floor plate, which originally in the eastern Pacific pointed north, might in the end have been swinging into the Indic Ocean from the south. These dates are further corroborated by the ocean floor chronology of the "Eocene Triangle" identified earlier—the "crotch" between the Ninety East Ridge and the edge of the Sunda Shelf, which was bent away from that Ninety East Ridge.

With Australia/New Zealand pulling north and veering east, the later Eocene would have been the time when the new "Southeast Indic Rift" would indeed have been squeezed into place, henceforth to split the Indic Ocean diagonally. The diagonal Southeast Indic Rift, which down south was established by an edge of Antarctic seafloor crust, came into being because Australia concurrently vacated that region.[38] **In the Indic Ocean, Jonathan Dehn was able to date this "mid-Eocene" tectonic event, tephrochronologically, at 42.7 Mya.**

Eureka! We now can see the Finger of Time point at the moment in tectonic evolution when the Sunda Shelf was bent eastward and away from its former flange, the Ninety East Ridge. This event could only have happened after the viscous asthenospheric "rubber-cord," that bound the tip of South America into the Bight of Australia, had been cut. At the moment of this snap, the upper asthenospheric belt, under the Sunda Shelf,

---

[38]Antarctica with its accreted triangular ocean plate, which derives from the northeastern Pacific, can best be seen on the NOAA Isochronic hemisphere, Figure 19, at the topographical hemisphere on Figure 16, and the author's Figure 6.

recoiled and loosened the entire Sunda crust for free travel. The rest is Southeast-Asian Pacific-Antarctic geologic history.

In global perspective, it now appears that a single process of Earth expansion has torn our planet's crust and spread its oceans, with accelerated spreading in the southern hemisphere. This same spreading has severed Africa, the Americas, and Southeast Asia like three petals of a flower which opened toward a southern sun. This southern expansion has united the three regional oceans—the Pacific, Atlantic, and Indic—into a single world ocean. From the blossom of the flower that opened has spun forth, counterclockwise, the round continent of Antarctica, to occupy the opening as if it had been a visiting bumblebee.

Of course, I make no pretense of knowing from where our planet received direct sunrays during the Eocene. Geologists and paleontologists (Prothero and others), have noticed the onset of glaciation in Antarctica during the Oligocene. They have concluded that the whole planet became colder then.[39]

Generalizations about this changing climate, for the entire Earth, often proceed as though Antarctica has remained stationary and can therefore be used as a reliable thermometer for the whole planet. In light of my explanations, however, the change in Antarctica's climate could have been the result of the continent's movement away from direct sunshine. But I hesitate to render a final opinion, because I do not know from whence that sunshine came or what the inclination of the planet's axis was in those times. Relative to the present globe, according to my perception, Antarctica would have been moving south

---

[39]Four essays on Eocene and Oligocene climatic events can be found in Donald R. Prothero and William A. Berggren, *Eocene-Oligocene Climatic and Biotic Evolution*, pages 131-217. Princeton University Press, 1992.

somewhat earlier than the epoch for which Prothero and Berggren have dated its cooling. We also do not know whether, as a result of the southward transition of continental mass the tilt of the planet's axis has been altered.

8

The Pacific Ocean

In order to go back to a smaller globe, with a single shell of continental crust and no deep oceans, most early terrella makers have simply subtracted the area of the oceans and then tried to rearrange Wegener's Pangaea as the original shell on a smaller sphere. They did not hesitate, arranging the continental fragments wherever they seemed to fit best. They were laboring under the spell of Wegener's *vagabund* paradigm—the notion that once upon a time the continents were *"wanderers"* who, somehow, came together to form the All-Land Pangaea. In the way that continents, in the beginning, drifted in from anywhere across the mythological All-Ocean Panthalassa, they also could disassemble and disperse again, or open up the Atlantic just as they pleased. Such free drifting seemed all the more reasonable on an expanding globe, where extra space was being created, continuously. Extra space increased the freedom of continental segments to wander.

Until recently all reduced globes, terrella models, were constructed as I myself did in 1979, as if to restore the ocean-less globe in a single-stage reduction. But thanks to worldwide ocean floor exploration, to drilling and magnetic profiling with ocean-going vessels, and thanks to a better topography and chronology of the planet's crust, this single-step approach is no longer sufficient. The more data one has, the more one must interpolate before taking a next step forward. Now we have a nearly complete ocean floor chronology. For relative dating, Earth-scientists have recorded horizontal patterns of magnetic reversals, starting along mid-ocean ridges. They also succeeded in establishing the magnetic profile of crustal strata in vertical drilling cores. The sedimentary layers were dated by paleontologists on the basis of fossil remains that were embedded in seafloors. Fossils, which at that time helped the ocean explorations to be successful, were frequently known beforehand, from the strata of ancient shallow seas which now can be found on dry land. Basement rock could be dated by measuring radioactivity.

This means that makers of terrella models now find themselves blessed, and simultaneously obligated, to match not only continental contours, but also to explain the isochrone sequences in the oceans of nearly 200 million years. The procedure of matching one continental shore with another, now requires an understanding of coherent sequences of ocean floor expansion between them, so that in addition to seeing ancient shorelines re-matched, temporal sequences can now also be traced, pictorially in marked striped space.

Whence came the Roundness of this Ocean?

The interpretation of isochrones is not a simple deductive procedure. Isochrones, along ocean floors, were magnetically imprinted not first in schoolbooks, but in nature, and into actual ocean floors of hardening lava while fresh ocean floor was being filled-in along spreading rifts. Before we attempt to read isochrones in the Pacific, it behooves us to step back a little ways from this ocean and raise the simplest preliminary questions that we are able to ask. From the distance, we notice that the Pacific Ocean is essentially round and furthermore, that the continent Antarctica is round as well. How do a round ocean and a round continent end up being neighbors? There must be some type of repeatable scientific experiment that could demonstrate probabilities and help answer this question in a more systematic manner.

Back in 1979, during my first reaction to Wegener's "continental drift," I performed a variety of experiments that would replicate Earth expansion on balloons which I inflated. I observed how slabs of putty behaved on an expanding and therefore flattening balloon surface. Photographs of these experiments are given above, in Figures 10-11. These images illustrate tensile folding, flanging, and expansion squeeze, as well as the severance of island chains along continental contours. In 1997 and 2002, to answer the question about the roundness of continents, I devised a different kind of balloon experiment (Figure 20).

Antarctica is the most round continent and the Pacific is our most round ocean. Though the continent is no longer located in the ocean's womb, and the size of the womb has been enlarged many times, the two entities together are still situated close enough to trace their original unity.

There seems to be a natural law and order at work that governs the formation of continental and oceanic contours. It is not my aim to define this natural law mathematically; rather, I hope to describe the process well enough so that it can be visualized and somehow demonstrated.

Even while doing my 1979 putty-and-balloon experiments I anticipated the possibility that a first continent, which is slowly being broken from the hard shell of an expanding planet, would likely be round. Back then, already, I had been reflecting on tensions that would want to hold a spherical crust together. The paradigm that guided me came from the unlikely realm of ornithology. Somewhere at a farm I have seen chicks hatch from their eggshells by cutting a circular opening with their beaks. I assumed that nature usually follows its own and easiest path.

Personally, I had no difficulty applying this logic to the shell of an expanding planet. The "eggshell breakage from within paradigm," in any case, seemed less arbitrary than the "continental *vagabund* paradigm" that Wegener has applied to explain the behavior of what, anyhow, would only have corresponded in my paradigm to "eggshell fragments after breakage." It seemed reasonable to me to let a growing chick in its shell, equipped with a beak, puncture from within to exemplify the role of an expanding force. While my own sense of logic was satisfied, it turned out that readers could not follow my paradigm-leap with ease. Their need for scientific experimental proof, regarding the shapes of first continents, called for an expanding force that could be demonstrated to be more obviously "blind" than a chick.

As my first step toward an analogous experiment I reintroduced my 1979 hypothesis, that the most natural form of a first continent, broken from the crust of a sphere to which expansion pressure is applied internally, would be a circle.

In 1997 an experiment was devised, repeated in 2002, which translates the "expanding eggshell" paradigm into a manageable "expanding balloon" paradigm. A video camera was immensely helpful for recording the process. Applied to a balloon, the problem emerged as follows: If an inflated balloon bursts, air pressure is released radially. The sudden localized outward rush of air pressure tears the balloon skin differently than a simulation of slow horizontal crustal tearing would, as for instance it would be expected by slow Earth expansion. While a planet's expansion pressure is radial as well, it is for the most part modified into horizontal tension by the planet's gravity and cohesion. Thus analogously, along the skin of my experimental balloon, the speed of bursting had to be slowed to a rate at which only slower horizontal tearing would occur.

Every scientific simulation has its limitations. The balloons available to me were not exactly spherical. They had the shape of a teardrop oval, and therefore only slightly more than half of the shape of a balloon could approximately represent an expanding sphere. But then, the teardrop oval shape of a balloon need not be seen as a disadvantage. Instead of true roundness, the working hypothesis may simply anticipate the possibility of a "teardrop oval" as its first continental fragment. To slow the breaking process, one balloon skin was slipped inside the experimental balloon to slow the radial force, and a transparent protective skin was slipped over the outside of both skins to restrict the experimental middle skin to horizontal tearing. The middle skin, that is, the experimental balloon, was punctured with a needle, so that it would start tearing sooner and less violently than the two untouched liners would.

Experiment One.—The middle balloon skin was pre-punctured exactly at the top, and it held up to an inflation diameter of about 16 inches. When it broke, the skin was divided into equal halves. The result was a truism. Two portions of a spherical crust, broken by expansion pressure into equal halves, necessarily do assume rounded shapes. This result was obtained several times when the balloon had been punctured exactly at the top. Inasmuch as this result can be envisaged easily enough, it need not be illustrated here.

Experiment Two.—The middle balloon skin was punctured about 20 degrees from the top center. This time the resulting portions were of unequal size. A smaller roundish "continental" patch was being peeled free, in relation to the remaining continental areas on our planet more or less proportionate with the size of Antarctica. Several attempts produced a continent with a genuine Antarctica tail! For typical images see "a," "b," and "c," in Figure 20. The presence of these tails is intriguing and

appears significant for understanding the formation of a continent like Antarctica as well as of the round Pacific cavity. Tails
were formed toward the end phase of the tear, rather than from
the starting point at the puncture. My balloon skin tore along a
curved path, almost symmetrically. But because the tension
and/or speed were not quite equal in both directions, the curved
path failed to complete a true circle. The distance between the
two paths, along the end runs, created the tail.

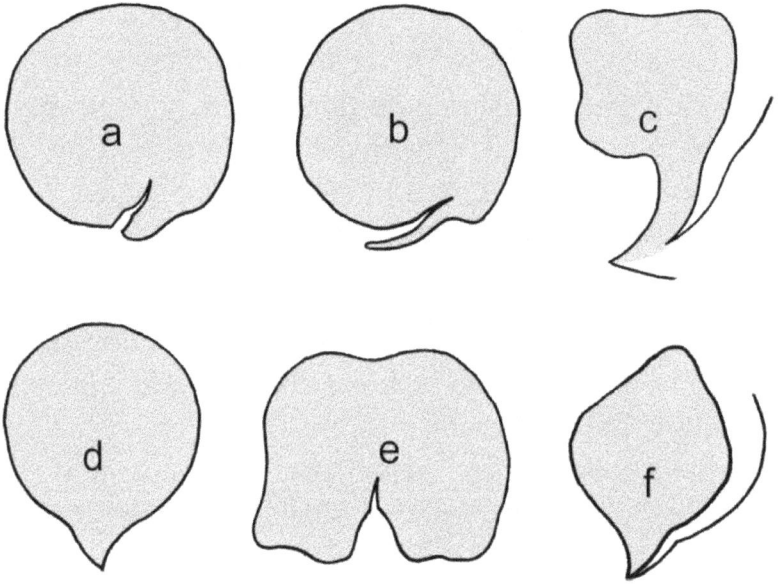

Fig. 20. Typical fragments from double and triple-balloon breaking experiments.

While the shape of an Antarctica, upon an expanding sphere,
is thereby clearly shown to be among the possibilities for first
continents, some variations can happen. If the two paths had
met as they did in images "d" and "f," the continent would not
sport a shrunken "mammalian" but perhaps more of a "reptilian"
tail. If the crust had begun tearing and then hesitated, the paths
would have reverted for a while, as in images "c," "e" and "f,"
and the continent would not have turned out to be very round.

If the tearing had changed direction and the curves had met at the other side with force, two stubs instead of a tail could have resulted, as in "e."

The observed variations do not negate the overall positive result of these balloon experiments. There does seem to be a natural pattern that determines the possible shapes, of first continents when they are being torn out of whole crusts of expanding spheres. If only one of these experiments had produced a shape similar to Antarctica, we would have to accept the likelihood that this continent was the first that was torn by expansion of the planet. This is not at all a question of experimental statistics. The basic reality is that Planet Earth indeed has one round continent with a tail. And it also has one expanded round cavity next to it in the shape of the Pacific Ocean, accentuated by a Ring of Fire. Insofar as our planet already has these features, we are not establishing the probability of obtaining them by a ratio of positive experimental results. In the case of our planet this probability is a posteriori, 100 percent.[40] Inasmuch as no other viable mechanism for these shapes has been suggested, the probability that Planet Expansion has created these features stands presently at one hundred percent as well.

The most natural form of a first continent torn from the crust of an expanding sphere, which generates and equalizes expansion pressure internally, does tend to be a circle. A more detailed description of the process, based on careful observation, is now possible. A round continent, breaking out from the

---

[40]A similar round-like "continent" can be found at the frosted North Pole of the planet Mars. Its placement at a pole, and the presence of round Antarctica at our South Pole, raises the question of whether poles determine the place at which round continents originate or are destined to be. The Martian coincidence suggests at least a working hypothesis. We may ask next whether the axis of our planet has tilted with the movement of Antarctica, or even whether the pole itself has moved together with Antarctica.

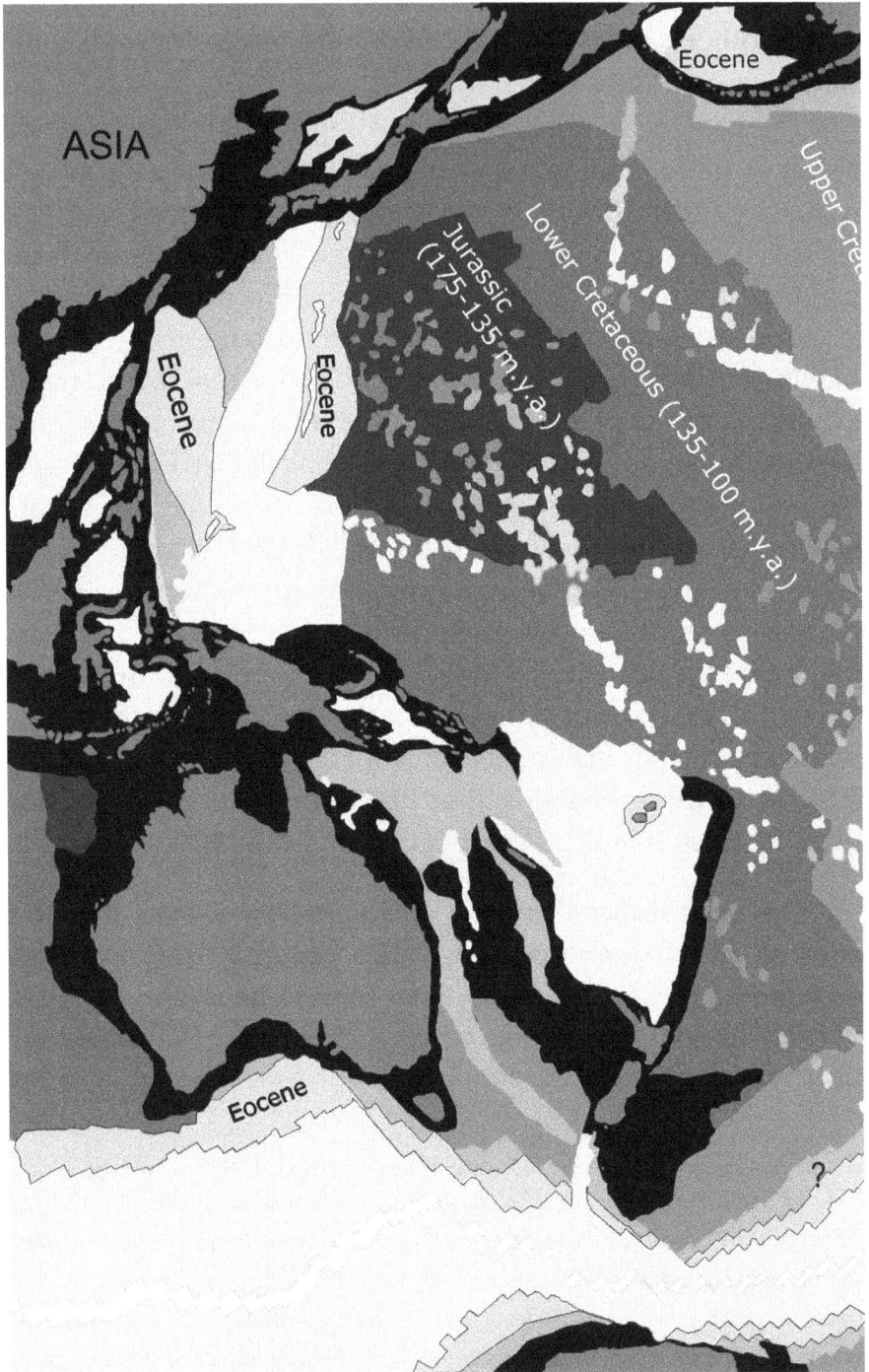

Fig. 21a Western Pacific, Drawn after UNESCO map 1988.

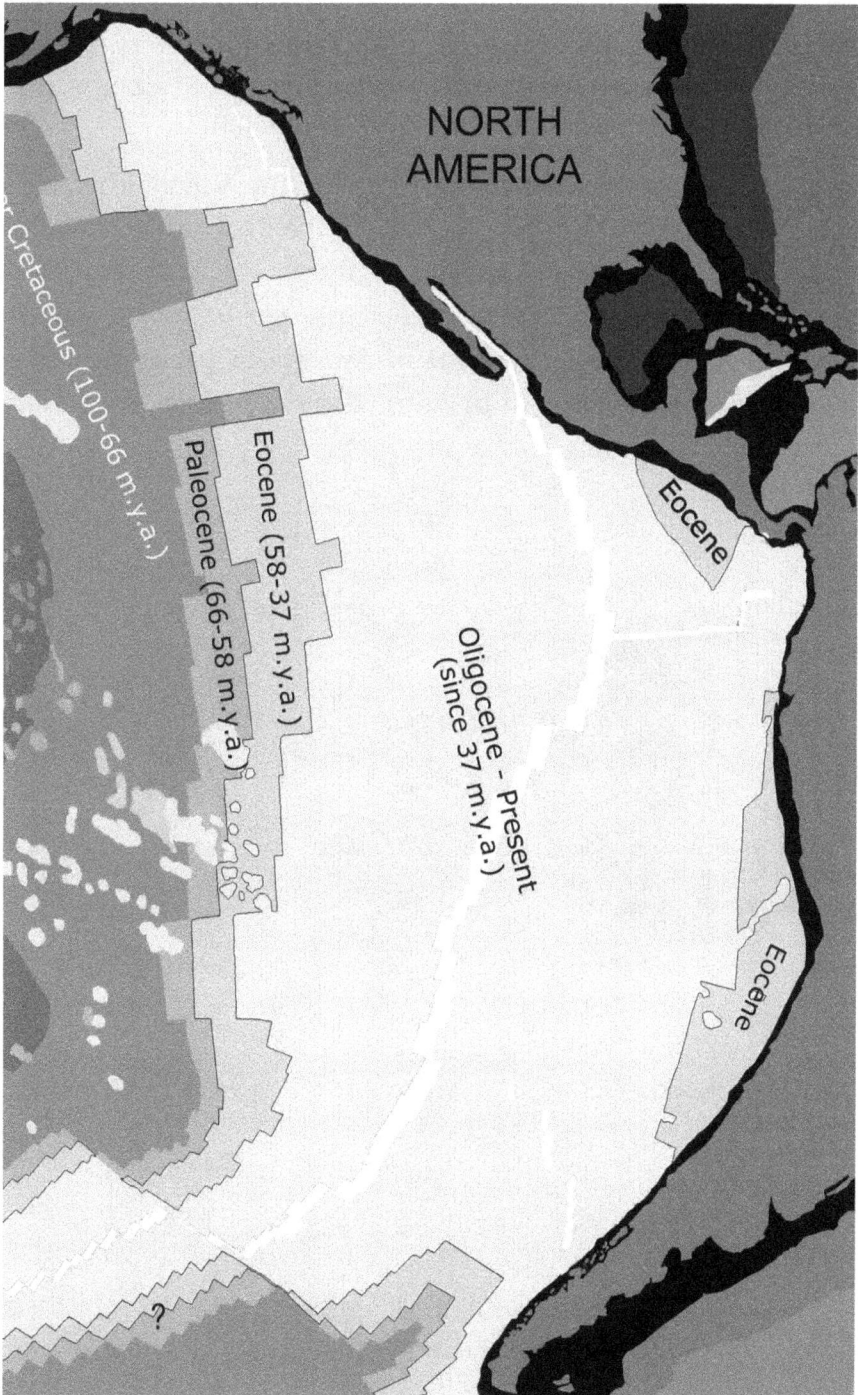

NORTH
AMERICA

Cretaceous (100-66 m.y.a.)

Eocene (58-37 m.y.a.)

Paleocene (66-58 m.y.a.)

Eocene

Oligocene - Present
(since 37 m.y.a.)

Eocene

?

Fig21b. Eastern Pacific. Drawn after UNESCO map, 1988.

crust of an expanding sphere, which on account of irregularities in crustal composition and cohesion, or for other reasons of unbalanced expansion-pressure, cannot complete a true circle, may manifest the differential in the form of a tail.

The oldest Jurassic ocean floor patch on our planet has been found in the western Pacific, where also the first round continent might have originated and come from, eastward. **A corresponding stretch of Jurassic ocean floor, albeit not yet very well investigated, is found along the edge of Antarctica precisely where, according to my hypothesis of Antarctica's extraction, such ocean floor could be.** (Figure 22)

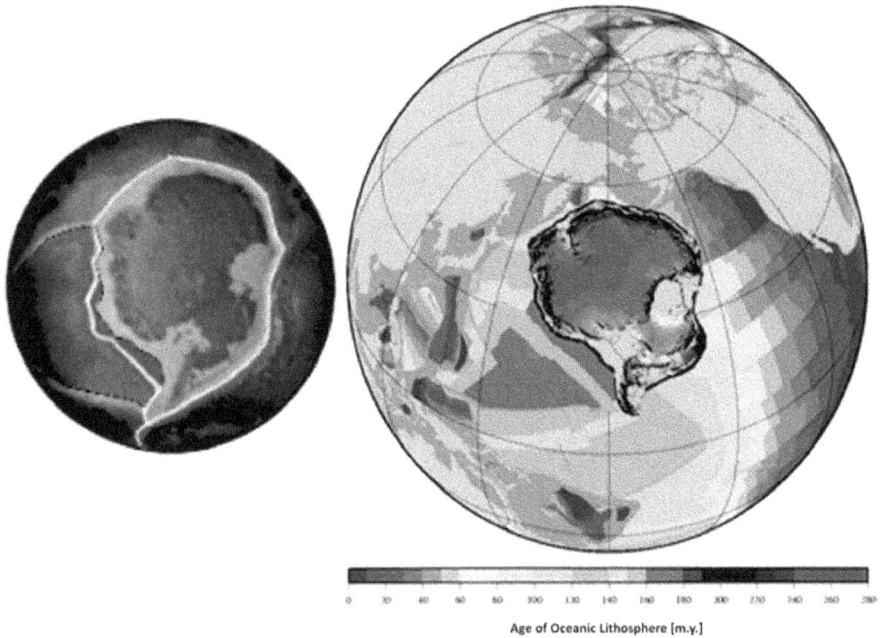

Age of Oceanic Lithosphere [m.y.]

Fig. 22. The Jurassic Pacific: a) zero ocean; b) Antarctica superimposed from NOAA Topo-hemisphere onto contemporary NOAA-Isochrone-hemisphere.

The Pacific Ocean can conveniently be divided into three parts—the older Northwest, the younger East, and the greatly irregular Southwest. The Pacific Ocean is essentially round, a fact that is emphasized by the presence of a seismically active

Ring of Fire. Admittedly, the Ring of Fire itself has been badly distorted and bent inward in the Southwest. To simplify explaining the Pacific here in a single chapter, the irregular intrusion of Southeast Asia, into the Southwest Pacific, has been explained in preparation already in the previous chapter on the Indic Ocean.

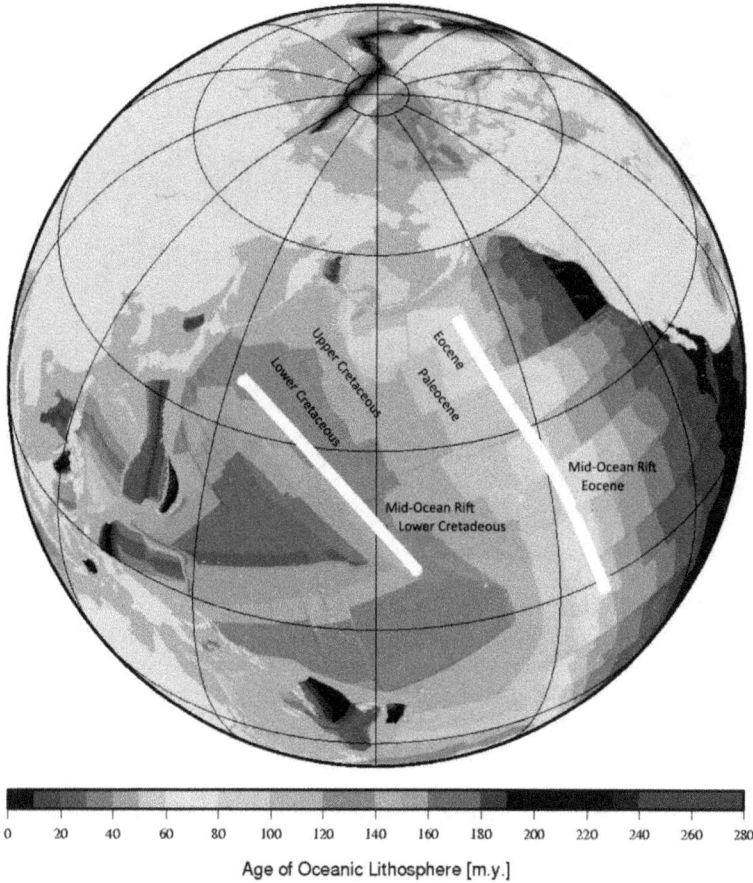

Fig. 23. Approximate Sequence of Pacific Mid-ocean Rift, Cretaceous until Eocene. Based on NOAA Isochrone-hemisphere.

I propose that the Ring of Fire, of strong seismic activity, outlines the expanded scar from which the continent Antarctica was carved over time. The round continent was torn loose, gradually, beginning by Jurassic ocean floor spreading in the western Pacific. Rifting continued afterward among Cretaceous and then among Paleocene ocean floors (Figure 23). At the beginning of the Eocene the round continent was seated in the eastern portion of the present Pacific. There the Antarctic Plate was still embraced by the western shores of the Americas.

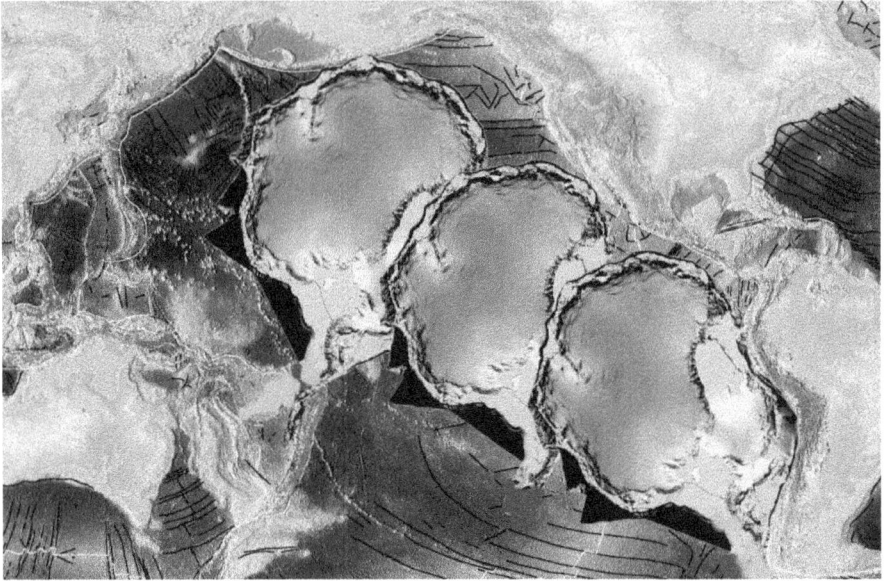

Fig. 24. Evolutionary Sketch Map, showing the movement of Antarctica in relation to South America and to the expanding Pacific—Cretaceous to mid-Eocene—on a present-day map. For Eocene ocean floor remnants see Figures 21a-21b. During the opening of the Pacific, along its eastern boundary, Antarctica was in the embrace of both Americas. As a result of the Eocene Transition, the inclinations of both Americas were radically altered.

Upon the smaller Earth, before and into the Jurassic epoch, land and shallow seas undulated up and down; the seas widened and shrunk at shallower depths. Then, beginning slowly in the Triassic (252-201 Mya) and in the Jurassic (201-145 Mya), a deeper set of oceans began to open up, by deeper rifting. At that initial stage, the spreading of deep oceans began by tearing at the coastlines of the future continental segments. The oldest Jurassic patch of ocean floor, in the western Pacific, initiated a separation that eventually would impact the contours of five continents—Asia, Antarctica, the two Americas, and Australia.

At that time New Zealand still had not separated from Australia, and Australia was still tightly seated at the cape of South America, cushioned in part by what was to become the tail off Antarctica.

The Jurassic ocean in the Northwest Pacific represents the first oceanic phase of continental separation on planet Earth. The starting date that is given for this process, in the *UNESCO Geological World Atlas* (1988), is 160 million years. A revised age of 175 million years (ODP, Leg 129) has been suggested for that same floor. This contrasts with 150 million years for the North Atlantic (ODP, Leg 149) and with still less for the Indic Ocean. My own zero-ocean terrella reconstruction of 1994, tracked the Jurassic patches that are found alongside Antarctica's Weddell Sea and Queen Maud Land to the northeastern edge of Jurassic floor in the western Pacific (Figure 22).

The first two Jurassic sides of the rift were separated, quite normally by Cretaceous spreading. The irregular edge of the world's oldest ocean corresponds in size approximately to the combined width of the Jurassic patches that still adhere to the rim of Antarctica. Similar steps of ocean reduction will reunite not only two Jurassic remnants from the Pacific and Antarctic oceans, but will also bring together pairs of Jurassic ocean floor which have become separated by Cretaceous spreading in the North Atlantic, as well as in the Indic Ocean. (Figures 6, 21, 22).

The oldest patches of ocean floor along the American coasts are Eocene, and the eastern portion of the round Pacific is its youngest part. Therefore, for the time being—until the chronology of the UNESCO Geological World Atlas of 1988 and drilling cores can be reviewed for implications relevant to my hypothesis—the Eocene epoch (58-37 Mya) will have to be postulated as the timespan of the great Eocene Transition. Antarctica's exit from the eastern Pacific, apparently happened later during the Eocene.

The flank of the Antarctic Plate that now faces south-eastern Africa was pre-cut along the Pacific's mid-ocean rift which, from its Jurassic beginnings up to the Eocene, has been spreading and moving eastward. **The western side of the Pacific Eocene rift is still present and traceable. As a strip of Eocene floor, it runs longitudinally up the middle of the Pacific.**

Please note, all isochrone stripes along the Central Pacific Rift have been seriously widened and elongated by the ocean's general expansion. Like the floors of earlier epochs, they have been widened by sections along transverse faults which appear distributed longitudinally quite loosely, and eastward.

For a long time after the departure of the Antarctic plate, from along the Eocene Mid-Pacific Rift, the western middle-aged floors have not encountered any obstacles to their eastward elongation and thinning. (see Figure 23)

Remnants of the Eocene rift which encircled Antarctica, along its east side, and which cut the round continent away from the coasts of the Americas, can still be found adhering to the coasts of Middle- and South-America (see Figure 21b). This last, continent-splitting segment of Eocene rifting, along the American shores, has finally severed the Antarctic Plate from its original American embrace.

The round continent was loosened during the Eocene, while Earth expansion manifested itself lopsidedly in the Pacific and the Southern hemisphere, and while the global belt was stretched. Its breaking enabled the Antarctic Plate to twist counterclockwise and slide south. Two Eocene half-rifts were left behind, bounding the eastern Pacific and attempting to join, ever so timidly, down the middle of the young Southeast Pacific through its still raw and wounded ocean floor.[41]

---

[41]The Eocene Transition, presented here as a hypothesis, explains why, according to Menard, the efforts at ocean floor drilling in the young "Southern Ocean" have yielded much ambiguous information.

The time it took for mid-regions of the eastern Pacific to harden is unknown time (*tempus incognita*). Henry W. Menard said early on, that spreading ridges in the eastern Pacific Ocean are ephemeral. The post-Eocene ridges which he sought are still difficult to detect in our days. Ridges could rise to become detectable only after new oceanic crust had evolved, cooled, and hardened to enable welding and mending to occur, along with concurrent magnetic encoding of the ridges that flank the spreading rift.

For comparison and contrast—the spreading rift in the Atlantic Ocean has been in place three times longer than the one that is presently taking shape in the young Eastern Pacific. The Atlantic rift began as a narrow, continent-splitting rift, whereas the eastern Pacific, during the Eocene, suddenly was bared of continental crust and needed to heal a huge raw asthenospheric surface. Seamounts alongside the Atlantic rift, nevertheless, could only have gotten uplifted after a certain width of the ocean floor curvature had spread, hardened, and then collapsed to provide Expansion Squeeze that would provide material weight and "hydraulic" pressure for uplifting the mounts.

In the Atlantic Ocean, the spreading rift is a central feature, and it should come as no surprise that the Lamont oceanographers, who concentrated on this ocean, were adamant at finding the Mid-Ocean Rift (MOR) as a universal feature in all the oceans. Those working along Southeast Pacific ocean floors, however, before they could start plotting mirrored magnetic isochrone reversals on maps, required far more patience to detect a central rift at which they could begin counting.

Antarctica, the round continent with a tail, is shaped like the number "9." That same 9-shape also happens to be the contour of the Pacific Ocean and its Ring of Fire. As a rock that is thrown into water broadcasts expansion rings from its point of impact, so the predestined shape, of the crust of the round continent torn from the shell of a sphere, is still replicated topographically along its Pacific Rim and by its defining Ring of Fire. Meanwhile, the Pacific Ocean has expanded to exceed the size of Antarctica's crust nearly twelve times.

The eastward moving 9-shape of Antarctica reverberates in magnetic anomalies which are embedded in hardening seafloor lavas of the Pacific. Cartographers have recognized the creation of ocean floors as isochrone sequences. Isochrons represent probable prior locations of Menard's "ephemeral" spreading rifts. They rendered visible the epochs that paleontologists had postulated earlier (Figures 23, 24). We now hold in our hands their grandest achievement—a nearly seamless chronology of the world's ocean floors.

Pioneers of the Plate Tectonics Revolution were also able to help improve ocean floor topography. Their earlier data, from depth soundings aboard ships, could more coherently be synthesized with the addition of satellite altimetry.

From the Jurassic through the Lower and Upper Cretaceous, through the Paleocene and into the Eocene—for the time during which Antarctica still occupied its place in the embrace of the Americas—the isochrones along the ocean floor repeatedly conformed to the pattern of a figure "9." This trend holds up even for the late Eocene, after the continent had departed and the southwestern region of the Pacific Ring had been severely indented and intruded upon by Southeast- and Austral-Asia. The new distortion did not destroy the "9" figure, but rather utilized it to enhance the implied oceanic shape.

The basic shape of the vacated Pacific Ocean has remained a circular cavity with room for a tail. Marvelously! Antarctica was born from the Pacific womb, as a Figure 9, almost as if it had been born from under its own tail.

In relation to the Eocene Tectonic Event, my choice of the term "distortion" is therefore quite negotiable. Switching the time perspective, one may ask the question another way: What, in the expansion history of the Pacific, has been the actual distortion? This round ocean does, after all, still conform to Antarctica's figure "9." Perhaps in this context we should also think of the circum-Pacific extremely-stretched global belt of continents, which reached its breaking point during the Eocene. This stretch-distortion preceded Southeast-Asia's northeastward sliding. The stretching was a prior and even a greater distortion. From the global evolutionary perspective, what now appears as a great distortion in the layout of Southeast- and Austral-Asia can realistically be seen as a tectonic rectification that required a hundred million years to resolve.

East Asian Marginal Seas

The marginal Seas of East Asia throw additional light on geological changes that happened during the Eocene. Making allowance for a Paleocene sliver in the Philippine Sea which had rifted earlier, and for an Oligocene expansion that was added later, Eocene spreads dominate also in those seas. The remaining East Asian marginal seas, which lie in the northern hemisphere, were opened quite uniformly during the Eocene. Such far-reaching Eocene uniformity, extending to the Sea of Japan and the Sea of Okhotsk, and continuing through the Bering Sea, cannot just be a minor coincidence. During the Eocene, East Asia abandoned huge flanges of coastland which became island chains in the western Pacific Ocean. The island chains lie anchored along old Benioff Zones and deep-cooled anchored trenches which the expanding Pacific has jarred. Cooling deep into the crust has added foundation bulges under the crust, down into the upper mantle (see Figures 10, 11, 14).

The formation of East Asia's marginal seas and island flanges can be understood in terms of the larger Eocene Tectonic Event, namely, as a general release of tension along the global continental belt. It happened when Australia and Southeast Asia snapped loose from the tip of South America.

How is such a sudden and widespread event even thinkable? Indeed, within the precincts of eastern Asia alone, these associations appear out of context. However, if one brings the remainder of Southeast Asia, the oceans and worldwide expansion stresses into the picture, then all appears coherent and obvious. In the global context we notice that, after the break of the continental belt, and the rifting of the Eocene seas, not much continental curvature was left beyond the East Asian mainland that could continue to hug the Pacific hemisphere.

"Global circumferential slippage" is a consequence of uneven horizontal cohesion and tension, generated over time unevenly by random Earth expansion while the continents were tearing apart.[42] With two thirds of the Planet's surface torn into, their cohesion among the remaining continental crusts became increasingly precarious. **After the loss of Antarctica from the embrace of the Americas, all continents of today still manage to hang together on the world map as a tattered curtain of continental crust. And that crust is now, for the most part, hugging the same hemisphere.**

With the Eocene Event, at the Pacific Ocean side, the waters of Panthalassa have indeed won their battle for hegemony over land. Continental shores around the Pacific will henceforth have no choice but to slip to and fro while widening their marginal seas. For an emergency measure, they have jarred their ocean

---

[42]The idea concerning "global circumferential slippage" in eastern Asia occurred while doing balloon experiments. Continental "skins" that cover only slightly more than one-half of a balloon, have a tendency to creep and to slip from the curvature of the expanding substratum.

trenches deeper, to cool and anchor their foundations. Trenches help avoid the danger of marginal slippage temporarily. While the planet's gravity will see to it that continental crusts do not fall off their hemisphere into space, the Pacific Ocean nevertheless has, for the foreseeable future, been destined to grow a little faster than the other oceans. As a result of its continental slippages, our planet has become a little more lopsided. Fish and humankind, having fins and submarines as their only recourses, are far too weak to influence their lopsided environment. So far humankind has even failed to balance its own societies and swarms of smaller lives.

With circumferential slippage and a general lack of continental cohesion, the lands on this planet will continue to leave deep-rooted flanges, and the seas will continue to crack open deep-cooled ocean trenches as boundary bulwarks for anchorage. The coastal ranges of eastern Asia had 130 Ma to build up tension against circumferential slippage. Some 43 Mya the continental belt lost its cohesion and tension. Whereas Asia lost its island chain, the Americas lost the largest flange amongst them all—the entire continent Antarctica—almost as if to pay homage to Alfred Wegener's *Wandernde Kontinente*. After the Eocene Event, the stress-gauge along the former global belt, in the south, was reset to zero.

Nevertheless, all the continents, and to a slight degree even careening Antarctica, still remain joined to their former neighboring continental crusts. Only South America and Australia have been completely severed from each other. Notwithstanding the Ural Mountains, Europe and Asia always formed but a single super-continent. From Gibraltar, along much of the western Mediterranean Sea, as well as along the Sea's eastern shores, Africa still is attached to the Eurasian continental landmass. Asia and North America still are nicely connected along the Bering Strait as well as around the Polar Sea, along northern

Greenland. Only a small rift in the North Atlantic, at Spitzbergen, cuts North America from Europe. The two Americas still are connected across Middle America, and Australia still is geologically part of Southeast Asia. On the topographical world-map, Antarctica appears to be the only fully liberated continent on our planet. But even this *vagabund* of a continent still drags its tail nostalgically from the direction of South America. Some scar tissue of that tail appears to be lodged, still, in rubble at the Cape of South America.

9

The Eocene Transition

Prothero and Berggren on the Eocene Transition

In the year 1992 Donald R. Prothero and William A. Berggren took a persuasive step in revising the timescale of the Eocene-Oligocene climatic and biotic evolution. The basic orientation of their work rests on the new chronological data that were collected in light of the new orientation of Plate Tectonics geology—the same direction from which this writer also attempts to obtain his data. There is one interpretational difference, however. While this author assumes Earth expansion as the primary key for understanding isochrone sequences, Prothero and Berggren work with what has become the dominant hypothesis worldwide. They recognize the expansion of ocean floors, but then, to preserve the notion of a steady-size planet, they consign all surplus ocean floor downward via convection currents, to a hot recycling process in the Earth's mantle.

All of us, even though we harbor different conclusions, are motivated by the goal of obtaining the most reliable sequential data. The question, of whether the interpretations of the data should hereafter be read on behalf of an expanding Planet Earth can safely be postponed for now. For the timespan of the Eocene, at least, both approaches may eventually find ways to enhance each other.

Prothero and Berggren think, and I agree, that "the transition from the Eocene to the Oligocene epoch was the most significant event in Earth history since the extinction of dinosaurs. When the Antarctic ice sheets appeared, major turnovers took place on the land and in the sea, eliminating forms adapted to a tropical world and replacing them with the ancestors of most of our modern animal and plant life...." The separation of Australia from Antarctica was a factor in changing the oceanic circulation patterns and the world climate.[43]

I happily recognize this revised timeline of postulating a gradual 10 Ma long Eocene-Oligocene extinction, with its major impact occurring in the 41-40 Mya range. On that account I am encouraged to introduce here again Jonathan Dehn's date of 42.7 Mya for a primary event at the Ninety East Ridge which, in my opinion, sent the Sunda Shelf, as well as Australasian lands and sea-floors, veering northeastward. It is my perception that Jonathan Dehn, in the Indian Ocean, actually dated part of something much bigger than mere upheaval along the Ninety East Ridge, namely, a tectonic twist that involved the Americas, Antarctica, as well as Australia.[44] The circum-polar southern continents have begun twisting apart and eastward, almost in unison.

Prothero and Berggren suggested that an increase in distance between Antarctica and Australia, a re-channeling of ocean currents, might have been responsible for climate change that gave Planet Earth a colder climate. This change required the evolution of more cold-resistant species.

---

[43]Donald R. Prothero and William A. Berggren. *Eocene-Oligocene Climatic and Biotic Evolution*. Princeton and Oxford: Princeton University Press, 1992, book cover and pp. 23f.

[44]The timespan between the Prothero-Berggren dates and those of Dehn is negotiable by the fact that the former have relied quite heavily on North American Eocene evidence. It could have taken some extra time before the full impact of Antarctic and Australian events might have affected life in North America. The events postulated by my hypothesis could equally well have filled most of the Prothero-Berggren ten-million-year window for the Eocene-Oligocene Extinction.

All this is wonderful and supportive embellishment for my Earth Expansion hypothesis. The southward twisting of Antarctica, and the severance of Australia which initiated the Eocene Transition, explain the icing of Antarctica and the cooling of the southern hemisphere much better than the unprovoked "wandering away" of Australia, that Wegenerian orthodoxy postulates. That suddenly, and for no geological reasons at all, a "precut" Australia could have drifted away, and left Antarctica (or Pangaea) without any asthenospheric horizontal cohesion to overcome, seems improbable.

Inasmuch as Antarctica has been chosen as the thermometer for Eocene- and Earth-history climate studies, we should at least know where this continent hails from and how cold it got before it arrived at the South Pole. This set of questions requires us to ask, where on Earth this continent was sitting, and when, and how over the course of its evolution it has moved. These questions, of course, only make sense with the proviso that all available data pertaining to subduction, to convection currents, Benioff zones, Earth-as-well-as-ocean expansion, and the Alaskan Earthquake of 1964, are revisited and re-evaluated with an intentionally relaxed and neutral engagement in chronological reasoning. For the time being, all such considerations should be approached as being still bracketed with doubt.

The immensely expanded Pacific cavity has remained true to Antarctica's continental shape. This happenstance implies a great amount of viscous coherence and resilience in the lithospheric substratum, possibly involving the entire depth of the asthenosphere. It indicates a physical tendency of "elastic" upper mantle materials, for expansion in response to an implied retention or "physical memory." Such could be propagated by changing nuclear and/or chemical configurations into something other or similar. We do not really know what all can happen under asthenospheric pressures and temperatures.

At any rate, some condition in the upper mantle has effected the preservation of the spiral 9-shape in the dynamics of the Pacific ocean floor expansion. I personally suspect that this type of elasticity would be inherent in the general viscosity of the middle and lower asthenosphere.

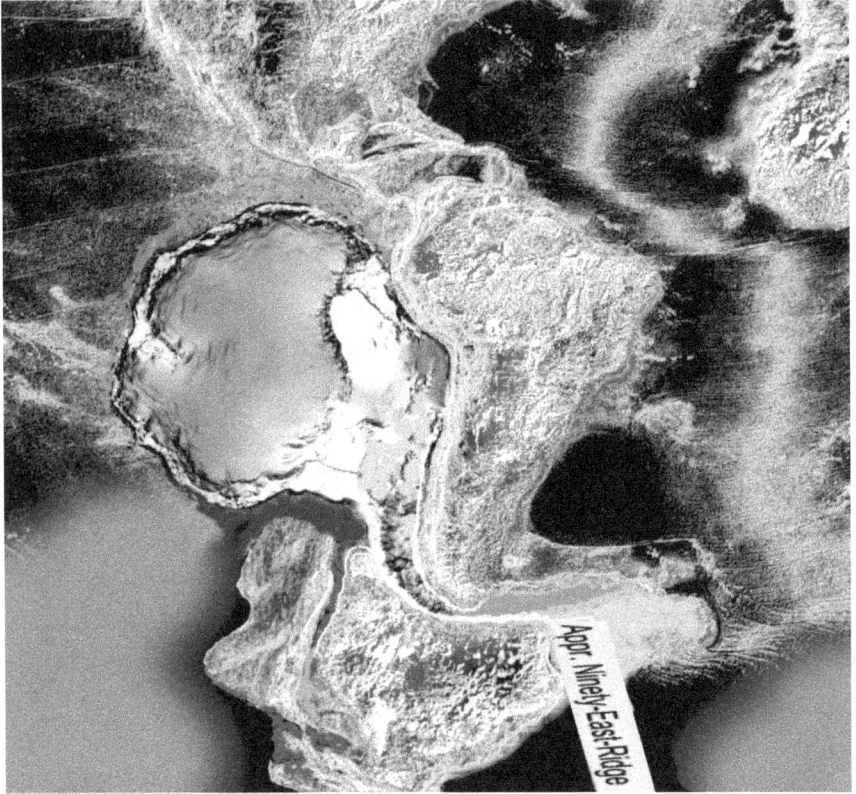

Fig. 25. Antarctica and Australia/New Zealand/New Guinea superimposed on a contemporary NOAA map. The Ninety East Ridge was lined up approximately with the Americas and the circum-global belt of continents.

When the Eocene Event commenced the northern wedge of ocean floor, which Antarctica had accreted since the Jurassic, swung counterclockwise through the eastern Pacific, by a westward curve. While the tail of Antarctica (which also may be visualized as a heel or toe) was wedging eastward, Australia was loosened from its seat at the cape of South America.

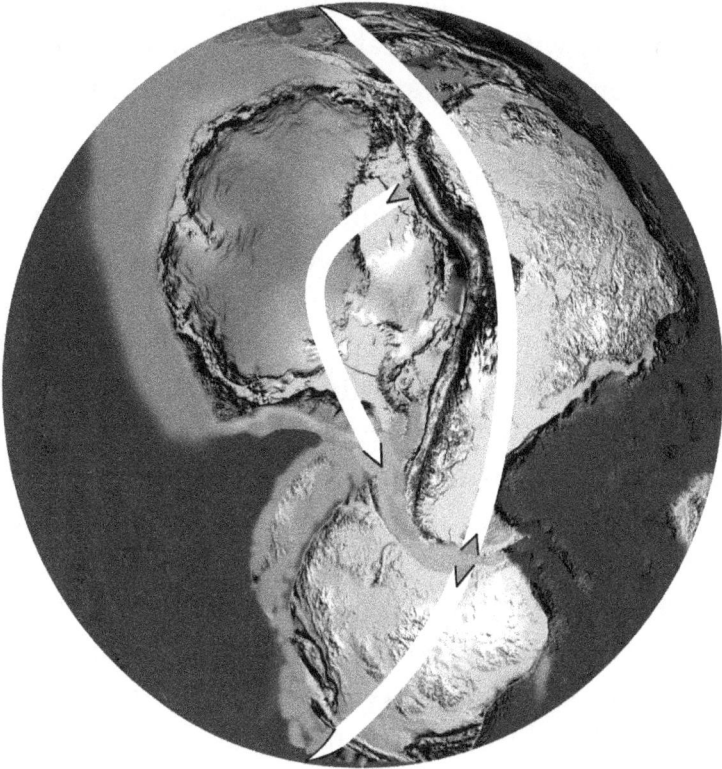

Fig. 26. Tensions and pressures along the Global Belt before the Eocene Event.

Antarctica, energized by its counterclockwise swing, has wedged its partly embedded tail into the widening gap, eastward. It scooped a portion of the South American cape ahead of itself. Eastward it scooped as far as the Sandwich Ridge, and when it withdrew from there, its pullback movement opened the Sandwich Trench (Figure 27).

When the separation of South America and Australia was complete, Antarctica continued to be carried onward by the velocity of its counter-clockwise swing. It continued its roll by bending its "wedge" abruptly south. The wedge was reshaped into a harmless tail. Antarctica continued its swing until it came to rest, almost dead center at the Planet's South Pole. (Figure 28)

Meanwhile, at the other side of the globe, Australia moved northeastward, with New Zealand up front, and with Jurassic and Cretaceous ocean floors and the Broken Ridge in tow. While so moving, Australia was part of the movement that twisted the Sunda Plate northeastward. Continental segments over half the globe appear to have been shifting. How could such an immense amount of movement get under way on a planet that had all its continents embedded among hardened Cretaceous and Paleocene ocean floor crust? It could happen because in its eastern Pacific area, considerably east of Southeast Asia, the huge Antarctic Plate was vacating its space. And before Antarctica could completely turn out of the Pacific, Earth expansion loosened up all continental crusts in the faster expanding southern hemisphere. It was movement on a broad scope.

The resulting distance between Australia and South America consists mostly of recent ocean floor. This means that at the beginning of the Oligocene epoch the Kermadec-Tonga Ridge was still closer to Australia and South America, and New Zealand has pushed older ocean floor against the back of Antarctica.

Aside from a small twist-eruption, around Fiji, there are no Eocene ocean floors indicated between the Paleocene and Oligocene spreads east of Australia. This means that Australia's eastward movement has absorbed all the Eocene expansion spread that could have accrued for ocean floors there. After the Australian Plate reached its northeastern limit, at the other side of the globe, it began to rebound westward.

Continental crusts are resting upon the asthenosphere, and when these crusts get jarred and nudged horizontally—as from expansion flattening or neighboring pressures or tensions—then, as a result of localized jarring, certain pockets in the upper asthenosphere will decompress. Viscous mafic rock becomes magma which then, in turn and at opportune moments, lubricates the bottoms of the crusts and enables them to leak or to slide. Such prerequisites to movement can get in place long in advance of actual movement, at various ocean depths.[45] Accordingly, it appears that at the Eocene Triangle, west of Sumatra some 42.7 Mya, a relatively "cool mechanical" event of loosening has occurred. Crusts so jarred, resting loosely on a sphere, lubricated by magma and sea-water, may turn curves more easily than they can slide linear. When Antarctica began to poke between South America and Australia, these similarly loosened crustal plates in the southern hemisphere, began slipping in unison.

Yes, the asthenospheric continental belt had been overstretched. It was pulled straight. Antarctica was nudged sideways by it, westward, and was expelled from the clasp and embrace of the Americas. The Middle American lands were thinned and pulled lengthwise by stretching and straightening. Their stretching and the implied transform-faulting sheared off Antarctica to be loosened and expelled westward. When Antarctia moved south, it stabbed its curvy wedge, or "heel," southeast-ward between the cape of South America and Australia. (Figures 26, 27)

---

[45]There appears to be considerable disagreement among geologists about exactly how much liquid magma the Planet has been able to produce. Enough rock-melt has been produced over time to weld and to lay down the magnetized ocean floors, two thirds of the Earth's surface, 5-10 kilometers thick. Some magma also appears to have been involved in the uplift of mountains and in most movements of continental crust. The question regarding the amount of magma appears to be a wrong question. If asthenospheric creep-rock can turn into magma as a result of decompression, then the resulting amount of available magma may vary greatly, and quickly. Another lubricant is sea water—a mineral-rich liquid with which the ocean basins are filled. The same forces that remodel ocean floors also seem to liberate seawater. But because we do not really understand the origin of water on this planet, we cannot include this substance in our theorizing without suffering some embarrassment. To suggest comets as sources of our Planet's water answers nothing. It merely kicks the question a little ways farther out into space.

The entire southern hemisphere was thereby loosened and twisted eastward rather easily. Moreover, this loosening happened in a pre-weakened region, where approximately 100 Ma earlier the cape of Africa had been torn loose from above the "toe" of South America. By the time all the mud had settled, and the waters cooled, Antarctica had slid into the place that Australia had been rifted away from. The Americas had straightened, whereas Southeast Asia had become severely twisted. Aside from Australia and Antarctica, other continents have changed their positions over considerable distances—Southeast Asia and Austral-Asia more so, and the Americas less.

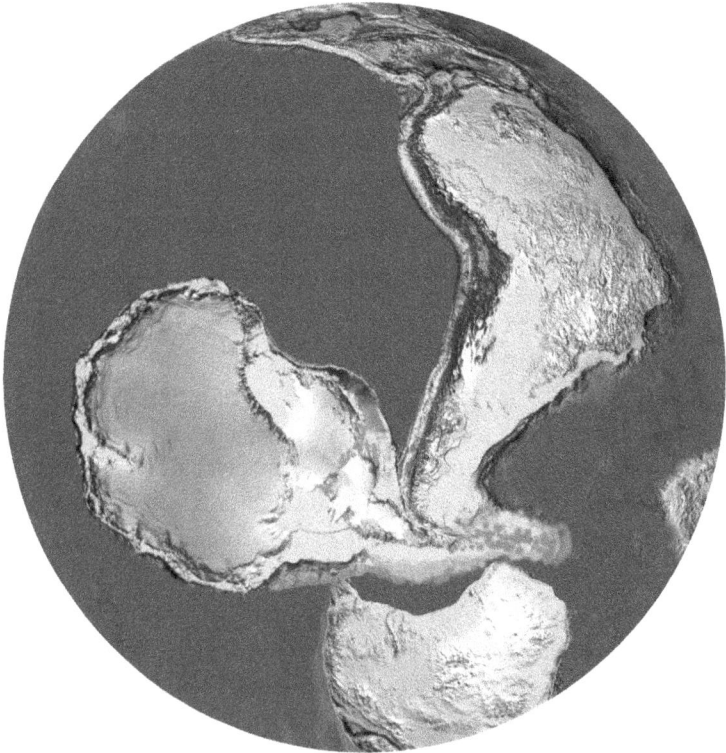

Fig. 27. The Eocene Transition in Progress, ca. 42.7 Mya.

The immense quantities of crust, having reseated themselves upon semi-elastic upper-mantle rock, upon a rotating wobbly planet, could not quiet down. Upon this expanding sphere, their adjustments have been reverberating all the while. There could be no absolute rest. We are being told that today the larger portion of the Australian continental collage is moving westward to tighten the conspicuous knot of islands at the Banda Sea, the Celebes Swirl. In the Andaman Sea some fragments of the old Sunda Flange, and a 9.2 earth-quake under a flange that is being abandoned, west of Sumatra, cracked in 2004, to bring

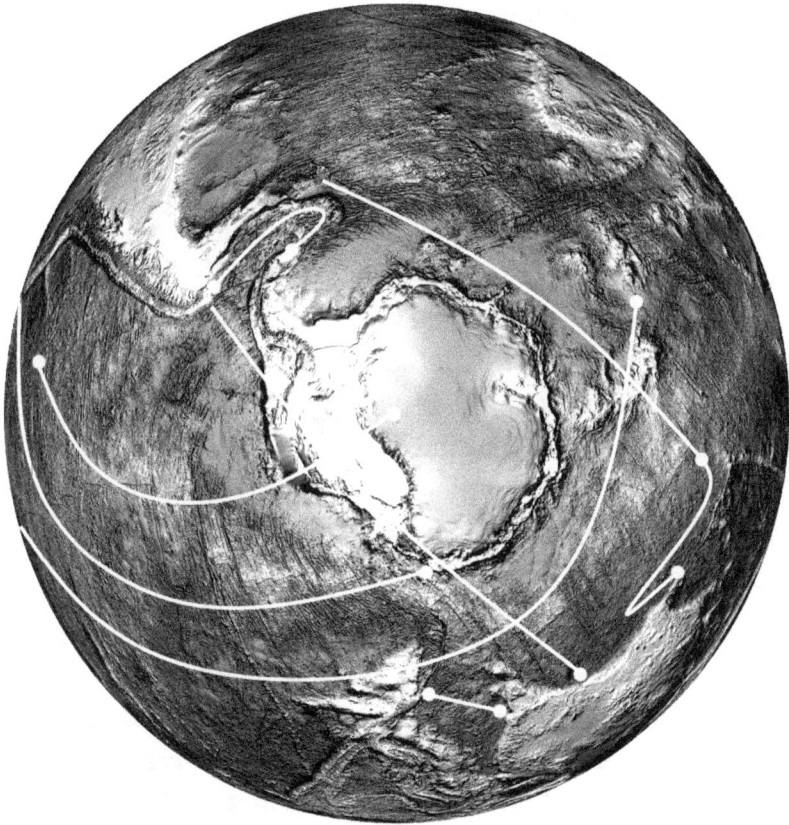

Fig. 28. Approximate distances of movement during the Eocene Transition; on a present-day NOAA Topo-map of the southern hemisphere. From its previous position (Figure 27), Australia was pulled away, for Antarctica to swing in. Then Australia with New Zealand, and Upper Cretaceous ocean floor, could drift east into vacated space. Thereafter, Australia began to ricochet westward.

on the deadliest tsunami of our time. Smaller ridges in the Andaman Sea were recorded as having moved to the Southeast. This means that today the Sunda Shelf is still not quite finished with its Eocene bending.

Cretaceous Floors in the new Ocean

Having arranged the Eocene Earth in relation to the present globe, roughly as in Figures 24-28, a seemingly difficult puzzle still awaits resolution. It pertains to the Eocene Tectonic event and, more precisely, to the ocean floors that now lie between New Zealand and Antarctica.

As far as the entire Antarctica Plate is concerned, it turned away counter-clockwise from the Americas, later during the Eocene. It was expelled from its place while North- and South America were stretched straight, longitudinally, by the then still asthenospheric coherent global belt of continents. All the while, Antarctica applied its velocity and mass to the movement of its "heel," to drive a wedge between Australia and the cape of South America. Australia, the "belt buckle," was loosened and pulled northeast-ward at the other side of the sphere and away from South America. At its own side of the sphere, South America ricocheted northeast-ward as well.

The strained belt of continents, when it broke, pulled back its ends and relaxed its length. Since Australia's loosening, the Eocene ocean floor triangle in the northeastern Indic was being vacated eastward. The tip of the Antarctic Plate—moving in part as "plate" and as "mudslide"—swung and scraped itself counter-clockwise all the way into the southern Indic.

All plates that interacted with this movement, quite expectedly helped destroy much of the younger Eocene floors which had been forming around them. One can assume that a considerable amount of magnetic ocean floor information was thereby rendered unreadable as well.

While South America was pushing northeastward against the Florida region, the southern portion of North America was pushed westward across its western coastal spreading rift, probably wiping out, as it went, slivers of Eocene floor east of that rift. Whatever Eocene floor fragments might have survived along Antarctica's original eastern shores, from along the North American coast, will probably remain indistinguishable from Eocene floors that awaited the careening continent when it arrived at the Indic rift.

As the NOAA-Isochrone hemisphere illustrates, Antarctica now has loaded at its back a riddle-some patch of Upper Cretaceous floor, trimmed with Paleocene framing (Figures 6, 21a-b, 29). It faces west and sits at a place where, according to the first step of our interpretation, nothing of that sort should be. To this Antarctica shoreline should adhere, at most, small remnants of Eocene floor that originated along the shores of the Americas.

Eocene patches along the American coasts are far more generously shown at the UNESCO-map (Figure 21a-b) than they are on the NOAA hemisphere (Figure 29). But nevertheless, the amount shown by NOAA is sufficient to endorse our interpretation regarding the severance of Antarctica there during the Eocene. The primary issue anywhere along the western shores of the Americas is that at least some Eocene ocean floors have formed and survived as indicators. They could only have formed after Antarctica had partially separated itself from the Americas and made room there for Eocene floors to harden. The puzzle for the moment remains: How have Upper Cretaceous ocean floors come to lay at the present western shores of Antarctica, bordered with a Paleocene frame?

We observe how this patch of Upper Cretaceous ocean floor, according to its size and inclination nicely matches the larger ocean floor, of identical age, west across the present younger area and spreading rift. One may surmise that both of these portions of Upper Cretaceous floor were pushed against the back of Antarctica together as a single unit, soon after or during the Eocene Event. The moving force behind them loosened the continental unit of Australia and Southeast Asia. (See the marked area, Figure 29). This means that the spreading rift, stemming from prior to the Eocene Event, was still actively burning down, as well as upward from deep in the asthenosphere. The process burned upward through this imported stretch of Upper Cretaceous seafloor by melting. A large area came so to be torch-cut in halves by hot magma. The new rift has reset the Isochrone Code of magnetized reversals.

On the Pacific maps, which back in 1997 I have based on UNESCO Geological World Atlas information, I marked the "Paleocene" edgings with question-marks—reproduced herein as Figures 21a-21b. Draftsmen of isochrone maps, who on the basis of their own data and drawings had not yet sufficiently considered the movements of Antarctica, or had not wondered about the shifting of Australia and Southeast Asia toward the northeast, could easily be inclined to read the space between my black lines (see Figure 29) as regular cartographer-friendly seafloor. Henry William Menard, however, gave sufficient hints, that oceanographers who worked on drilling ships in the South Pacific regularly encountered difficulties while, all along, they were looking to find less ambiguous data.

My two black lines were therefore, during the late Eocene, still a single line which could have indicated a deep asthenospheric hot line of sorts, either of an earlier spreading rift or of a new rift that tried to average the former eastern and western coastal rifts of Antarctica. All this goes with also recognizing that in the course of the Eocene Event, South America has ricocheted a certain distance northeast-ward. In any case, the isochrones that appeared readable next to younger sea-floor crusts seem to belong neither to the Paleocene nor to the Eocene.

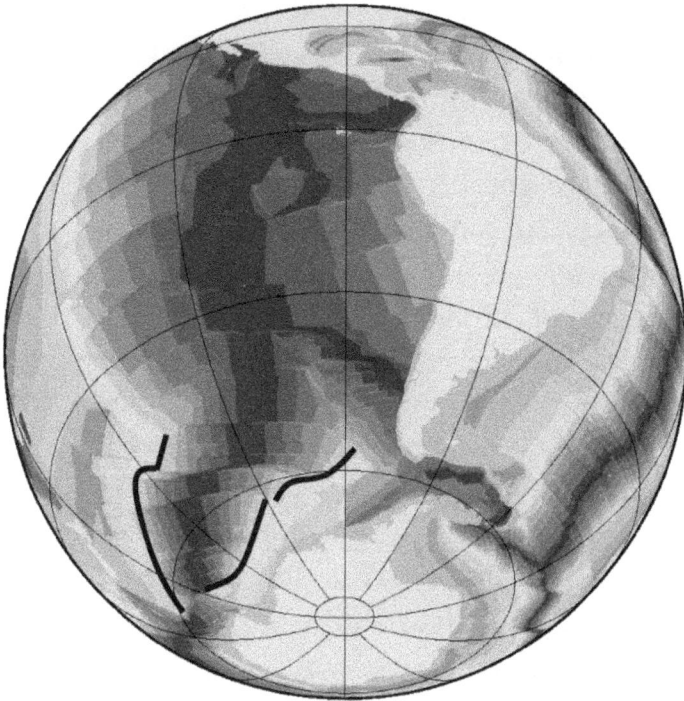

Fig. 29. NOAA Isochrone-hemisphere, shows Upper Cretaceous floors from the South Pacific, at the back of Antarctica, subsequently burned through by rifting heat from below. Black lines were added by the author.

The geographical hints in Figure 28 are clear enough. Australia and South America still face each other across the continent of Antarctica which, as anyone who looks at its curvy tail can see, has played the role of an interloper at the moment when those two continents became estranged. Thus, an answer to our final puzzle lies within reach.

As an immediate result of the Eocene Event, after South America and Australia had both recoiled northeast—at opposite sides upon the sphere—a larger area of Upper Cretaceous ocean floor (stronger and thicker than the still soft Eocene floors) were pushed eastward against the back side of Antarctica which, meanwhile, far to the south had come to face west-ward (see Figure 6). This arrival of older Cretaceous floors happened after the Antarctic plate had passed through the area. As a result of repeat-occupation between New Zealand and Antarctica, it seemed, not much young Paleocene and Eocene ocean floor crust could have survived. However, as soon as enough ocean floor was cooled and hardened, by sea-water, the formation of magma underneath could be slowed.

Before a young crust can restrain and contain magma, its surface needs to cool and harden. A thin ocean crust, like the thicker continental crust, needs to adjust to the flattening curvature of the expanding Planet, which means episodically it needs to collapse at its domed middle. The oceanic spreading rift, melting deep into the asthenosphere, is layering fresh magma unto the crust from below and is sealing with pillow-lava overhead. Within the rift a hefty bead of lava-weld is injected, to be shared by both flanks along the rift at the ocean floor halves.

As a new spreading rift is formed, the ocean floor is divided into two domed half-plates. From their collapsing domes both halves do squeeze surplus magma toward their flanks along the rift. To some extent the squeezed magma feeds the rift which draws most of its magma locally, from the asthenosphere right beneath the rift. Along the flanks of the rift, some of the sideways squeezed materials will go into building seamounts.

Slightly pre-cooled bulges of magma arrive and accumulate under the rims of the two half-plates at either side of the new spreading rift. In time, these bulges uplift chains of seamounts. It happens as a result of Earth expansion, because the two half-plates divided by the rift are flattening and squeezing their surplus magma sideways. In time the injected fillings will harden and will be strong enough to stand as rising peaks of seamounts. Today two bulging ridges along both sides of the mid-ocean spreading rift do connect all our oceans to form a functioning world ocean. In all oceans, the same natural processes of spreading are in effect.

The broad bulges of seamounts and their centrally running spreading rifts function in a manner analogous to the synclines that grow mountain ranges at the rims of continents. High continental plains correspond tectonically to the wide and still younger dome formations along seafloors. Mid-ocean Spreading Rifts correspond to virgin mountain-diapirs on continents. Wherever in the asthenosphere there is sufficient space, decompressed rock materials will become magma. And magma, when squeezed, will seek to create mountains or sea floors. The process of mountain formation, everywhere on land and in seas, requires domed plates to collapse, crusts to flatten, and magma to be squeezed sideways. Some magmas intrude into crevices as plutons and others are pasted to the foundations from beneath. They are all cooled from above. Then, in response to cratons settling, the mountain ranges rise by isostasy and by sustained added pressures from episodic Expansion Squeeze.

The rift that spreads the softened East Pacific now runs behind the Baja Peninsula into the Gulf of California which it has created. From there it dives under the State of California. During the Eocene this rift-line was active between the continents—west in the sea between Antarctica and the Americas. Now it still severs the Baja Peninsula from the mainland of Mexico.

Later during the Eocene, after South America was pushing against the southeastern corner of the North American landmass, the southern portion of North America went west-ward. Specifically, California was pushed across the former coastal rift-line. The land-features of the Rift are known as the San Andreas Fault. At the southwestern corner of the State of Oregon, this former coastal spreading rift appears again in the Pacific Ocean, apparently to continue along the ocean floor where formerly it has torched a fault-line deep into the asthenosphere to complete the carving of Antarctica.

Summary of the Eocene Tectonic Event

Inferring from the magnetic isochrone-based ocean floor chronology, the tectonics of the Eocene Transition Event can now be summarized. The southern hemisphere lay expanded, poised for several continents to be shifted simultaneously.

(1) The oceans, especially those in the southern hemisphere, needed to have expanded sufficiently for the Eocene Transition to happen. There needed to be sufficient young and soft seafloor spaces that could become wiggle room between the continents, to allow initial movement.

(2) Slowly and as part of the continental circum-global belt, the Americas were stretched and straightened, longitudinally. Antarctica was thereby nudged sideways and pushed westward.

(3) Antarctica was loosened and twisted counter-clockwise, out of the embrace of the Americas and continued moving by the Planet's rotational agitation and centrifugal force. Its heel was applied like a chisel to carve a rift along the Australian Great Bight. It scraped across the asthenospheric underlay of the global belt, similar to a glass-cutter, and then bulldozed some continental crust at the toe of South America, to scoop as far east as the Sandwich Ridge.

(4) The global belt was cut across in one swipe, and Australia was sliced away from South America, approximately at the place that during the early Cretateous had gotten weakened, where Africa was torn loose. Now South America and Australia pulled apart, northeast-ward, at opposite sides of the sphere.

(5) While South America recoiled north-eastward, it pushed against the southeast of North America. It nudged that continent westward, a little ways over the former coastal spreading rift which now, under land, has become known as the San Andreas Fault.

(6) A triangular plate of seafloor, attached to Antarctica's original northern edge, provided additional weight, length and velocity for the combined plate to swerve all the way into the southern Indic. Centrifugal force and the wobble of the rotating Planet has all the while agitated the extreme vertical incline of the asthenospherically loosened Antarctica. The continent of Antarctica proceeded to occupy the South Pole, a location which, on a rotating planet, is by far the safest place for a loosened round continent to be.

(7) Australia with its Broken Ridge and the entire Sunda crust were bent out of the way, northeastward. Antarctica's accreted triangular seafloor plate that once had touched southern Alaska was fitted into the southern Indic Ocean.

(8) While Australia vacated its space for Antarctica to occupy, Australia together with New Zealand pushed a large area of Cretaceous seafloor against the "back" of Antarctica. Said backside of Antarctica, prior to its southward swing, had been facing the eastern loop of its Eocene spreading rift which ran between itself and the western shores of the Americas. Some patches of Cretateous floor of what was pushed there can be found still hanging there today.

Epilogue pertaining to the Fifth Dimension

Karl W. Luckert—Added for the Second Printing, in May 2016

A hint was given already in the Preface, that the worldview of this writer was shaped between two languages, while moving from a German into an English environment. Wegener's Continental Drift hypothesis traveled the same road. However, Plate Tectonics science took shape in America and traveled outward in the opposite direction. The fertile Romantic "drift" within the German language, over the past two centuries, can be illustrated with the evolution of a song—a *Wanderlied*, or hiking song.

Tectonics of a Hiking Song.—Wilhelm Müller, a poet of Dessau, Germany, in the year 1818 has written a round of lyric poems, titled *"Die schöne Müllerin."* Franz Schubert, in 1823, has worked this bundle of twenty poems into a musical and has made it famous. A simplified version of the opening song, "Das Wandern ist des Müllers Lust," was arranged for men's choirs by Carl Friedrich Zöllner, in 1844. Arguably, this song has become the best known German folksong for millions of people.

Any person who went hiking could feel inspired by the four-voice Zöllner chorus, which loudly glorified the obligatory *Wanderschaft* of craftsmen. When an apprentice in a craft graduated, he was officially declared a "free fellow"(*Geselle*)—as this author himself was *"freigesprochen"* in 1952. The fellow could then go wandering another three years and work for several masters, far away from his home area. This was required before he could become a master of his craft. Academicians also travelled the road, but their journeys were not registered or controlled by guilds.

The educational effects of Zöllner's choruses were incidental, but they conveyed all the usual sentiments of the Romantic Period—the metaphors and hyperboles of love, longing, freedom, happiness, suffering, as well as contemplations of death and the prospect of eternal rest. For example, the miller-*Geselle*, in the Müller lyrics, appears driven by his love of a beautiful miller's daughter. When that love was rejected, he succumbed to self-pity and drowned himself in the mill-creek—or at least thought about doing so.

To orient Alfred Wegener's book, *"Die Entstehung der Kontinente und Ozeane"*(1915) in an evolutionary context, within Germany, we should focus specifically on the Fourth Stanza, below. There reverberates the sentiment of regarding stones as fellow wanderers. This metaphor elucidates why Wegener could have reflected on continents as wanderers.

Rushing mountain streams, rolling and thundering pebble stones or boulders, were actually seen only by those wanderers who braved heavy thunderstorms in the mountains. But with the help of music, the masses of ordinary folk could dream about wandering alongside wandering stones, which then would dance and run ahead of them. All laborers would have loved to wander more, and work less.

2. Vom Wasser haben wir's gelernt,
vom Wasser haben wir's gelernt, vom Wasser!
Das hat nicht Rast bei Tag und Nacht,
ist stets auf Wanderschaft bedacht,
ist stets auf Wanderschaft bedacht, das Wasser.

3. Das sehn wir auch den Rädern an,
das sehn wir auch den Rädern an, den Rädern!
Die gar nicht gerne stille stehn,
und sich mein Tag nicht müde drehn,
und sich mein Tag nicht müde drehn, die Räder.

4. Die Steine selbst, so schwer sie sind,
die Steine selbst, so schwer sie sind, die Steine!
Sie tanzen mit den muntern Reihn
und wollen gar noch schneller sein,
und wollen gar noch schneller sein, die Steine.

5. O Wandern, Wandern, meine Lust,
o Wandern, Wandern, meine Lust, o Wandern!
Herr Meister und Frau Meisterin,
laßt mich in Frieden weiterziehn,
laßt mich in Frieden weiterziehn, und wandern.

Fig. 30. Dora and Karl on their visit to the Great Bight, September 2006

The contemplated behavior of stones, for many German readers, belonged to the virtual reality of physics and nature, as well as of whatever knowledge could be acquired under the influence of music. These were the people whom Alfred Wegener addressed with his theory regarding drifting and "wandering" continents.

Almost a century had passed since Wilhelm Müller penned his words about dancing stones, and thereby fused the attributes and behavior patterns of stones with those of human wanderers. In 1912 some of these words enabled Alfred Wegener to think and communicate about continents that could indeed be wandering. And wandering stones and wandering continents did make some sense when they were contemplated against the background of music that already had inspired most everyone in Germany.

As Carl Friedrich Zöllner has amplified Müller's and Schubert's impact perhaps a thousandfold, by way of organizing men's choirs, so Alfred Wegener has enlarged Müllers wandering stones to the size of continents, to fit his world map. It was henceforth possible to contemplate, in the German language, the continents as being wanderers. The Romantic marvel, of stones traveling in water, became the science of Plate Tectonics.

It seems quite inappropriate to answer the Müller-Schubert-Zöllner-Wegener group with only the latterday scientific language of Plate Tectonics. I am aware that added poetry will not change the history of science. But for the sake of perspectives on possibilities, one should wonder what type of balancing notions, during the Late Romantic Period, could possibly have slowed down the linguistic drift in German, toward drifting continents.

I must honor Wilhelm Müller by way of responding in his own style. But the stones on which I dare sit, nowadays, no longer are of the dancing or wandering variety. But by replacing three stanzas of Müller's overture, I might just also avoid wanting to raise mountains by collision—or to keep something from growing larger by way of trying to subduct its skin.

> 3. Der Müller geht den Weg entlang
> den Wegener erbauet han, die Wegener.
> Der Wegner ruht auf Steinen aus.
> Er räumt sie sich vom Weges Lauf,
> Er räumt sie sich vom Weges Lauf, zum Bauen.
>
> 4. Die Kontinente reisen nicht,
> doch um die Schollen reißen sich, die Meere.
> Die Erd erfüllt sich innerlich,
> Mit Brunst und Feuer weiten sich.
> Mit Brunst und Feuer weiten sich, die Meere.
>
> 5. O Wandern, Wandern, hin zum Ziel,
> o Wandern, Wandern, hin zum Ziel, zum Bleiben.
> Der Wan-derer kann nicht mehr gehn.
> Am Grab ein Stein bleibt bei ihm stehn.
> Am Grab ein Stein bleibt bei ihm stehn. Er bleibet.

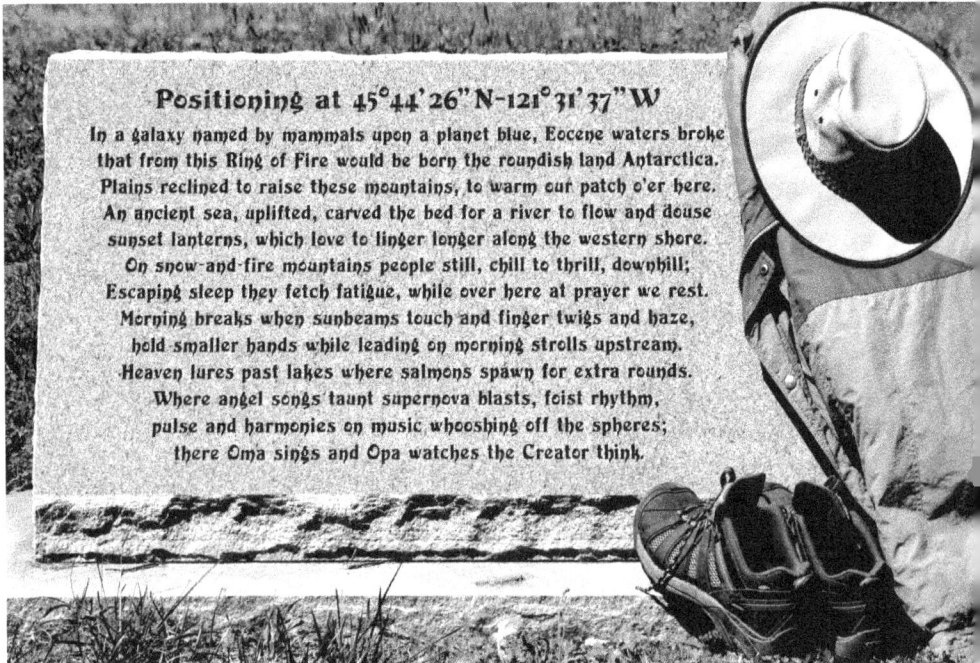

Positioning at 45°44'26"N-121°31'37"W

In a galaxy named by mammals upon a planet blue, Eocene waters broke
that from this Ring of Fire would be born the roundish land Antarctica.
Plains reclined to raise these mountains, to warm our patch o'er here.
An ancient sea, uplifted, carved the bed for a river to flow and douse
sunset lanterns, which love to linger longer along the western shore.
On snow-and-fire mountains people still, chill to thrill, downhill;
Escaping sleep they fetch fatigue, while over here at prayer we rest.
Morning breaks when sunbeams touch and finger twigs and haze,
hold smaller hands while leading on morning strolls upstream.
Heaven lures past lakes where salmons spawn for extra rounds.
Where angel songs taunt supernova blasts, foist rhythm,
pulse and harmonies on music whooshing off the spheres;
there Oma sings and Opa watches the Creator think.

Bibliography

Alvarez, Luis W. *T. Rex and the Crater of Doom*, Princeton: Princeton University Press, 1997.

Bevis, Michael and Barton Payne. "A New Palaeozoic Reconstruction of Antarctica, Australia, and South America," in Carey, S. W. *Expanding Earth Symposium*, Sydney, 1981, pages 207-213.

California Institute of Technology Tectonics Observatory: at http://www.tectonics.caltech.edu/outreach/ highlights/sumatra/ why.html. "Why Earthquakes and Tsunamis Occur in the Sumatra Region." "What Happened During the 2004 Sumatra Earthquake." "Using Coral to Track the History of Earthquakes." "Rethinking the Causes of Giant Earthquakes."

Carey, S. Warren. *Theories of the Earth and the Universe: A History of Dogma in the Earth Sciences.* Stanford: Stanford University Press, 1988.

_____. ed. *The Expanding Earth, a Symposium.* Sydney: Earth Resources Foundation, 1981.

Choubert, G. and Faure-Muret, *Geological World Atlas.* Paris: UNESCO, 1976-1988.

Dehn, Jonathan. www.aist.go.jp/GSJ/~jdehn/research/diss.htm.

Grand, Stephen P. and Rob C. Van der Hilst, and Sri Widiyantoro, "Global Seismic Tomography, a Snapshot of Convection in the Earth," in *GSA Today*, April 1997.

Hoshino, Michihei. *The Expanding Earth: Evidence, Causes, and Effects.* Tokyo: Tokai University Press, 1998.

Hsü, Kenneth J. *Challenger at Sea: a Ship that Revolutionized Earth Science.* Princeton: Princeton University Press, 1992.

Jacob, Karl-Heinz. "Erdexpansion—verkannte Geowissenschaftliche Theorie?" *Erzmetall*, 54 (10), 2001; pp. 473-484.

Jacob, Karl-Heinz and Scalera, Giancarlo, editors. *Why Expanding Earth? A book in honor of Ott Christoph Hilgenberg.* Technische Universität Berlin and Instituto Nazionale di Geofisica e Vulcanologia, 2003.

Lamb, Simon and David Sington. *Earth Story, the Shaping of Our World.* Princeton: Princeton University Press, 1998.

Luckert, Karl W. *Mother Earth Once Was a Girl: a Scientific Theory on the Expansion of Planet Earth.* Flagstaff: The Museum of Northern Arizona Press, 1979.

_____. "A Unified Theory of Earth Expansion, Pacific Evacuation and Orogenesis," in *Theophrastus' Contributions to Advanced Studies in Geology,* pages 61-73. Athens, Greece: Theophrastus Publications, S.A., 1996.

_____. "Expansion Tectonics, a video program." Part One: the Formation of Oceans; Part Two: the Formation of Mountains; Part Three: Story of Discovery. VHS and PAL, 84 minutes. Springfield, MO (VideoScript at <www.triplehood.com>), 1996.

_____. *Planet Earth Expanding and the Eocene Tectonic Event.* A Lufa and Triplehood publication. Also posted at www.triplehood.com, 1999.

_____. Plate Tectonics is Expansion Tectonics: the Tectonics of Rising Mountains and Growing Continental Plates. Bilingual DVD in English and German. Produced for the Theuern Conference on Earth Expansion, Germany, May 2003. Script downloadable at www.triplehood.com.

_____. "Four Theories of Earth Expansion and the Eocene Event." DVD. Produced for the "New Concepts in Global Tectonics" Conference, Urbino, Italy, August 28, 2004. Script downloadable at www.triplehood.com.

Maxlow, James. *Global Expansion Tectonics: Small Earth Modelling of an Exponentially Expanding Earth.* Glen Forrest, Australia: Terrella Consultants, 1996.

Menard. H. W. *The Ocean of Truth: a Personal History of Global Tectonics.* Princeton: Princeton University Press, 1986.

Meyerhoff, Arthur A. "Surge Tectonics evolution of southeastern Asia: a geohydrodynamics approach," in *Journal of Southeast Asian Earth Sciences.* Vol. 12, No 3-4, pp. 145-247, 1995.

NASA Jet Propulsion Laboratory and **National Geographic Society.** *The World Satellite Map,* 1998.

NOAA, GEMCO Paris, Geol. Survey Commission of Canada, Scripps Institution of Oceanography, Univ. of Sydney, Univ. of Texas. *Age of the Ocean Floor.* US Dept. of Commerce, National Geophysical Data Center, 1996.

Ocean Drilling Project: www.odp.tamu.edu/sciops/LegSummaries.

Parker, Sybil P. ed. *Dictionary of Earth Science.* New York: McGraw-Hill, 1997.

Pflafker, George. Henry C. Berg, ed. *The Geology of Alaska* (The Geology of North America, Vol. G-1). Boulder: Geological Society of America, 1994.

Prothero, Donald R. "Cracking earth and crackpot ideas," in *Skeptic Magazine,* vol. 18, number 1, 2013. Also at www.donaldprothero.com/files/92370160.pdf.

Prothero, Donald R. and **William A. Berggren.** *Eocene-Oligocene Climatic and Biotic Evolution.* Princeton: Princeton University Press, 1992.

Scalera, Giancarlo. *Variable Radius Cartography – Birth and Perspectives of a New Experimental Discipline,* INGV – Istituto Nazionale di Geofisica e Vulcanologia. Roma, Italy, 2013.

Scalera, Giancarlo and **Karl-Heinz Jacob,** editors. *Why Expanding Earth? A book in honor of Ott Christoph Hilgenberg.* Technische Universität Berlin and Instituto Nazionale di Geofisica e Vulcanologia, 2003.

Schatzman, Evry. *Our Expanding Universe.* New York: McGraw-Hill, 1992.

Smith Walter H.F. and **David T. Sandwell.** *NOAA Map*: "Global Seafloor Topography from Satellite Altimetry," 1997,

Strutinski, Carl. "Some Reflections on the Charts of the Ocean Floor: do they hide more than they reveal?" at www.dinox.org-publications-Strutinski2015.pdf.

Suzuki, Yasumoto and **Takashi Mitsunashi, Kisaburo Kodama, Yoshijiro Shinada, Seiki Yamauchi, Atsushi Urabe,** Boso Peninsula: *Guidebook of the Boso Peninsula.* International Symposium on New Concepts in Global Tectonics. Tsukuba, Japan, 1998.

Truempy, Rudolf. "Penninic-Austroalpine Boundary in the Swiss Alps: A presumed former Continental Margin and its Problems." *American Journal of Science,* Volume 275-A, 1975. pp.209-238.

Van der Hilst, Rob C., Sri Widiyantoro, and **E. R. Engdahl,** "Evidence for Deep Mantle Circulation from Global Tomography," in *Nature,* vol. 386, 10 April 1997.

Vogel, Klaus. "The Expansion of the Earth, an Alternative Model to the Plate Tectonics Theory," in *Critical Aspects of the Plate Tectonics Theory,* II, 19-34. Athens, Greece: Theophrastus Publications, S.A., 1990.

Wegener, Alfred. John Biram transl. *The Origin of Continents and Oceans.* New York: Dover Publications, 1966. Original publication in German, 1915.

Wertenbaker, William. *The Floor of the Sea, Maurice Ewing and the Search to Understand the Earth.* Boston: Little, Brown, and Co., 1974.

Yano, Takao, ed. *Proceedings of International Symposium on New Concepts in Global Tectonics,* Tsukuba 1998.

Subject Index

ISBN 978-0-9839072-6-8

www.ingramcontent.com/pod-product-compliance
Lightning Source LLC
Chambersburg PA
CBHW020155200326
41521CB00006B/378